AI Projects with
Raspberry Pi

AI Projects with Raspberry Pi
by Lucy Hattersley and Brian Jepson
ISBN: 978-1-916868-42-7
Copyright © 2026 Raspberry Pi Ltd
Published by Raspberry Pi Ltd, 194 Science Park, Cambridge, CB4 0AB

Raspberry Pi Ireland Ltd, 25 North Wall Quay, Dublin 1, D01 H104
compliance@raspberrypi.com

Editors: Brian Jepson and Lucy Hattersley
Interior Designer: Sara Parodi
Production: Brian Jepson
Photographer: Brian O Halloran
Illustrator: Sam Alder
Graphics Editor: Natalie Turner
Publishing Director: Brian Jepson
Head of Design: Jack Willis
CEO: Eben Upton

July 2026: First Edition

Table of Contents

Welcome

Artificial Intelligence (AI) is the buzziest of tech topics. You can hardly pick up a new piece of hardware or software without encountering an agent or model.

We are starting to see AI technology seriously being deployed across industries around the world. Businesses and governments are augmenting their workforce with AI agents, factories and industrial environments are using AI inferencing to check and manage the build process, and coders are increasingly using AI tools and workflows. Even artists and writers are pushing back, and moving forward, with AI generated content.

If you are interested in AI on a practical level, now is your time to shine. There are huge rewards for the engineers, coders, and makers who demonstrate an understanding of how AI concepts work and how AI technology can be deployed at the edge (on a local device) and positioned in the real world.

Raspberry Pi has developed a powerful AI hardware stack that can run AI models at real-time speeds without requiring a cloud service. This has powerful benefits for speed, security, efficiency, and reliability.

The mental rewards are almost as good as the physical. This stuff is right at the cutting edge of modern computing, and it is fascinating to get an understanding of where computing is, and where it is going.

In this book we explain how Raspberry Pi's hardware can be put to use with modern AI concepts, such as image and video analysis, text generation and translation, voice detection and speech synthesis, real world sensing and data analysis. We look at how to use models, and more importantly, how to customise and train models so they suit your needs.

About the authors

Lucy Hattersley is a technology writer specialising in programming, hardware, and maker culture. She is Editor of *Raspberry Pi Official Magazine* where she commissions, edits, and shapes articles for one of the UK's most beloved hobbyist publications. She writes tutorials with hands-on expertise in AI hardware, machine learning, and deploying local language models. An advocate for privacy-conscious technology, she also volunteers with FoodCycle in London and helps organise the Brockley Max community arts festival. In quieter moments she can be found reading actual books and wrestling with a cryptic crossword.

Brian Jepson manages the development, production, and distribution of books and magazines for Raspberry Pi Press, the publishing imprint of Raspberry Pi Ltd. With over 30 years of experience in the publishing industry, Brian pursues a passion for making computing and technology accessible to people of all ages and backgrounds. Brian also enjoys creating electronic devices that interact with the physical world. Outside of work, Brian collaborates with like-minded artists, technologists, and non-profits in Rhode Island to create experiences that get people excited about STEM/STEAM careers and hobbies.

Colophon

Raspberry Pi is on a mission to put high-performance, low-cost, general-purpose computing platforms in the hands of enthusiasts and engineers all over the world.

Since 2012, we've been designing single-board and modular computers, built on the Arm architecture, and running the Linux operating system. Whether you're an educator looking to excite the next generation of computer scientists; an enthusiast searching for inspiration for your next project; or an OEM who needs a proven rock-solid foundation for your next generation of smart products, there's a Raspberry Pi computer for you.

That's not all we do. For lower-power and real-time applications, Raspberry Pi Pico boards, and the growing RP2 family of Raspberry Pi silicon products, bring our signature values of high performance, low cost, and ease of use to the microcontroller space. Our high-quality accessories ensure you get the best performance from your Raspberry Pi products. Raspberry Pi publications provide support and inspiration to makers and professionals alike.

Every Raspberry Pi product is backed by extensive technical documentation, broad global regulatory compliance, a worldwide network of trusted resellers, distributors and design partners, and a global community of educators, enthusiasts and professional design engineers. If you want to add intelligence to your latest project, or latest product, there's someone in the Raspberry Pi community who can help make it happen.

Raspberry Pi is an engine for creativity, learning and innovation. With over 60 million of our computers sold in the last decade, there has never been a more exciting time to build on Raspberry Pi.

Raspberry Pi Press

rpimag.co/books

Raspberry Pi Press is your essential bookshelf for computing, gaming, and hands-on making. We are the publishing imprint of Raspberry Pi Ltd. From building a PC to building a cabinet, discover your passion, learn new skills, and make awesome stuff with our extensive range of books and monthly magazine.

Raspberry Pi Official Magazine

magazine.raspberrypi.com

Raspberry Pi Official Magazine is written for the Raspberry Pi community. It's packed with Raspberry Pi-themed projects, computing and electronics tutorials, how-to guides, and the latest community news and events.

Example code

Visit this book's GitHub repository for example code and other information, including errata: **rpimag.co/aiproj-resources**. To download the example code as a zip file or to check it out locally with `git`, click the button labelled *Code*.

If you've found what you believe is a mistake or error in the book, please let us know by using our errata submission form at **rpimag.co/aiproj-errata**.

Chapter 1

Practical AI
with Raspberry Pi

A complex technology on a compact computer

Artificial intelligence is an umbrella term that is rather vague. It covers everything from very specific technologies, such as machine learning, deep learning, and natural language processing through to the use of AI technologies in hardware and software projects. But it's also sometimes used as a science fiction plot device that pits human intelligence against perceived computer intelligence.

It's for this reason that many AI experts use the term artificial intelligence carefully and tend to home in on detailed technologies with distinct meanings such as machine learning or natural language processing.

This book will show you many practical applications of AI that work on Raspberry Pi hardware. There are several broad categories of AI applications that the book focuses on: computer vision, predictive modelling, voice, linguistics, and large language models (LLMs).

Because Raspberry Pi allows you to create physical computing projects, you can connect sensors and actuators to your Raspberry Pi and write code to interact with them. So, for example, you could use an AI model to learn gestures by learning from a sensor such as an accelerometer. You do this by collecting a large amount of sensors readings while you wave the accelerometer around. Then you provide additional data (a *label*) that indicates when the gesture occurred. Both datasets (the gestures and the label) go into the training process.

Then, when it's time to use the model, you could write some code that uses your AI model to detect whether the gesture is occurring, and if it is, you could take action by telling an actuator (such as an LED or motor) to do something. This book includes a project that does exactly that, but you'll also learn a wide range of other applied AI techniques.

Why AI?

AI is a polarising topic. As this book is being written (by humans!), the publishing industry is in the midst of a $1.5 billion settlement between Anthropic and a group of authors (**rpimag.co/anthropic-authors**). At the centre of this settlement is the allegation that Anthropic used the content of digitised books, to train their Claude model without compensation to the authors. (Full disclosure: one of the authors, Brian Jepson, is among the authors who may benefit from the settlement).

The settlement was reached without Anthropic admitting wrongdoing, but the fact that it happened, and involved such large sums of money, indicates that there is no broad consensus when it comes to the ethics of training artificial intelligence models. What's more, there really is no ethical framework for other aspects of AI. For one, industry and governments are building energy-hungry data centres while some feel that the environmental impacts are being completely ignored. Also, conversations around the ethics of where and how AI should be applied, particularly in the medical field, are in their infancy.

Learning roadmap

We're not going to argue that AI is an inevitability, and you should simply embrace it. But we are going to argue that you should study it and understand it. Among the most controversial members of the family of AI models are large language models (LLMs). They require large amounts of training data, large amounts of energy, and are black boxes that humans can't really inspect or fully understand. They only reveal themselves through the responses they generate, but these responses don't truly tell users anything about their internal state.

There are many AI models that aren't as fraught as LLMs. They are the workhorses of data science, and you'll learn about them in this book. For example, in Chapter 3, *An introduction to predictive models*, you'll learn how to create some of the simpler kinds of machine learning models, regression models. And although the final project in Chapter 7, *Speech to text and back again* involves an LLM, the bulk of the chapter uses models that are less complex, designed for specific rather than general purposes, and are trained on open source or public domain data sets.

Before we get into those models, however, we're going to start out in this chapter with some basic definitions and a quick overview of one of the foundation data structures (tensors) used in AI and machine learning. In Chapter 2, *The Raspberry Pi AI stack*, we'll step back for a moment and talk about the Raspberry Pi AI stack (software and hardware) and then get onto regression models in the chapter after that. Chapter 4, *Generate images from prompts* gets hands on (and a little fun), as you learn to install models that can generate images from textual prompts.

We'll then dive into how to run an LLM on your Raspberry Pi in Chapter 5, *Run a local LLM*, using both the Raspberry Pi's GPU and the Raspberry Pi AI HAT+ 2 accelerator. We'll then take a little detour into the world of magic in Chapter 6, *Training models on sensor data*, where you'll train a model to recognise gestures and build a magic wand that sparkles when you flick it just right.

Chapter 7, *Speech to text and back again* introduces you to all sorts of things you can do with voice: you'll learn how to transcribe spoken messages, respond to spoken commands, and even generate speech from text. And if that's not enough, you'll also see how to translate from one language to another and even create a voice-powered chatbot.

Finally, Chapter 9, *Recognise video and image content on the CPU* takes us back to the world of images, where you'll learn how to recognise and classify objects in images or video streams using the Raspberry Pi AI Camera or AI HAT+ (version 1 or 2). We have a lot to teach you before we can get to that, though, so let's get started.

From training to inference

One term we've used a bit already is *model*. An AI model is a representation of a system that learns from data. There are two distinct stages to a model's lifecycle: *training* and *inference*:

Training

During training, an AI model is taught to recognise patterns in data. The computer goes through the data, adjusting parameters (known as *weights* and *biases*) to improve the accuracy of the output. The training process is hugely computationally intensive. While you can experiment with it at home, it typically runs on dedicated hardware or within cloud datacentres.

There are two main types of learning, *supervised* and *unsupervised*. The type of learning imposes some requirements on the data you use for training. In supervised learning, data is typically *labelled*, which means that every combination of input variables in your data has a corresponding output value (the label). Those input variables are *predictors*, as they predict a particular value for the label. In unsupervised learning, you don't have labels, so the model needs to look for patterns in the data on its own.

Supervised learning is generally used to create models that need to predict a particular output based on certain inputs. For example, you might use supervised learning to create a model that can recognise specific objects, such as people, cats, and teddy bears. To do this, you'd collect many photographs, manually create a dataset that labels each object of interest (along with the coordinates of a square that indicates where the object appears in the photo). The training process would use the photographs and labels to create your model.

Unsupervised learning is good for models that look for patterns in data, such as identifying clusters of similar items in your dataset. For example, you might have thousands of comments from a forum or product reviews. After some data preparation, you could train a model that will segment the comments into several clusters: two, ten, one hundred (whatever you specify). The model will then be able to tell you which comments belong to which cluster, and it's then up to you to decide what to call each cluster (funny, rude, snarky, helpful, etc.). If you find that the model isn't segmenting comments into meaningful clusters, you can try training with a different number of clusters or tuning some of the training parameters

TRAIN/TEST SPLITS

In supervised learning, you will often need to set aside some portion of your training data as *test data*. After the training process has built a model, it will then use the test data to tell you how well the model performs. This works because the model wasn't trained on the test data, but also because the correct result is known for the test data.

In some cases, the training process will use an additional subset of your training data as *validation data*. In this case, the training process repeats for multiple iterations (*epochs*). For each epoch, the training process adjusts some parameters and uses the validation data to determine whether the model has improved. Typically, training repeats until it reaches a set number of epochs. It can also stop when the model no longer improves with each successive epoch. This is called *early stopping* and helps to avoid *overfitting* (when a model's results correspond so closely to the training data as to be useless with real-world data).

Inference

This is the process of running the model on new data (inputs that are not part of the train or test dataset). You pass the raw data (such as an image, text, or audio) into the AI model and get output in the form of a prediction, classification, or decision.

Taking the supervised learning object recognition model mentioned earlier, you could ask the model to identify objects in pictures that it's never seen before. It would then respond with its inference: the type of object (person, cat, teddy bear), and the coordinates of a square that surrounds the object.

In the case of the unsupervised learning example, you could submit a new comment to the model, and it would respond with the cluster (typically a number from one to the number of clusters) that the comment (probably) belongs to. We say "probably" because we are talking about models, after all: the quality of a model depends on having enough data but also having data that is *clean* (mostly free from errors, complete, and accurate).

Reach the TOP

While models can be extremely large, many are functional, albeit a bit slow, on a Raspberry Pi 5 without additional accelerator hardware. Many language translation models will easily fit in available memory and can translate text relatively quickly. For example, we translated the first two paragraphs of this section on a Raspberry Pi 500 in 10 seconds without any accelerator hardware (see "Machine translation" on page 151 for more details). Other kinds of models will run painfully slow, however. For example, generating an image from a prompt (Chapter 4, *Generate images from prompts*) could take hours.

With the addition of the AI HAT+ or AI Camera, you can run models optimised for those devices quite fast. These devices feature neural network accelerators for vastly speeding up inference time. This hardware enables you to perform real-time object recognition, for example.

It is possible to train some neural networks on Raspberry Pi, but the process is quite slow and it's more commonplace to train models on dedicated hardware and deploy them on computers like Raspberry Pi.

The AI performance of computers and accelerators is measured in *TOPS* (tera operations per second). This is the standard speed with which AI equipment is measured, and 'tera' is short for 10^{12} or 1 trillion.

Training and running AI models is hugely computationally intensive and relies on mathematical operations like matrix multiplication, convolutions, and derivatives. One TOPS is a trillion of these operations per second.

▸ **Coral USB Accelerator (4 TOPS)** — Early AI Raspberry Pi endeavours like Google's Coral accelerator were capable of around 4 TOPS.

▸ **AI Kit (13 TOPS)** — Raspberry Pi's AI Kit consists of an M.2 HAT+ with a Hailo AI acceleration module. New customers are now recommended to get the AI HAT+ instead.

▸ **AI HAT+ (26 TOPs or 13 TOPS)** — Two models are available, with Hailo-8 or Hailo-8L accelerator offering 26 TOPS or 13 TOPS inferencing performance, respectively. The Raspberry Pi AI HAT+ is fully integrated into Raspberry Pi's camera software stack, taking advantage of the neural network accelerator to run post-processing tasks such as object detection, image segmentation, and pose estimation.

▸ **AI HAT+ 2 (40 TOPS).** — The Raspberry Pi AI HAT+ 2 brings generative AI capability to Raspberry Pi 5. With 8GB of dedicated on-board RAM, the AI HAT+ 2 is ideal for running large language models (LLMs) and vision-language models (VLMs) locally and securely.

▸ **Raspberry Pi (around 0.5 TOPS)** — You don't need an AI accelerator to run AI models. A bare Raspberry Pi can handle AI workloads, just a bit more slowly. Although CPUs and AI accelerators are not directly comparable, you can expect around one half TOPS with a Raspberry Pi 5. (Your cooling setup has a big impact on actual performance.) If you connect custom hardware such as the Raspberry Pi AI Camera, you can perform real-time object classification at 30 frames per second (FPS) without using an additional accelerator.

Frameworks

TensorFlow and PyTorch are two deep learning frameworks. These provide extensive code support for developing and deploying models. The two frameworks have pros and cons, and both are worth investigating. There are many subtle differences between TensorFlow and PyTorch, but in general:

- ▸ **TensorFlow** — Better for industrial deployment. Supports Python, JavaScript, C++, and Java (**tensorflow.org**)

- ▸ **PyTorch** — Better for research and experimentation. Supports Python and C++ (**pytorch.org**).

Get to know tensors

Key to how artificial intelligence operates are *tensors*. A tensor is a multi-dimensional array of numbers. Think of it as a list with nested lists inside. These form a shape based on the dimensions:

0D Tensor (Scalar)

This is a single number, such as 8 or 3.14.

1D Tensor (Vector)

A list of numbers, like: [1, 2, 3]

2D Tensor (Matrix)

A 2-dimensional array, like: [[1, 2, 3], [4, 5, 6]]. That tensor's shape would be described as (2, 3).

3D Tensor

An array of matrices. This would have a shape described by the matrix references, like: (size, height, width).

ND Tensor

These are higher-dimensional arrays used in machine learning models. These are harder to visualise as they move beyond three-dimensions. For example, a video might have: (batch, frames, height, width, channels).

An AI model consists of layers of tensors. Imagine an array of these multi-dimensional arrays with links between them (that are weighted to adjust strength) and biases (that are used to adjust the output to fit). The shape of the network and the adjustments of the weights and biases are used to train the neural network to the task.

Running the models requires some intense mathematical processing, typically using matrix multiplication (for details on how that works, see **rpimag.co/matrixmult**). If you want to learn more about tensors and experiment with them, here are two great guides:

- TensorFlow: Tensors (**rpimag.co/tensorflowbasics**)

- PyTorch: What is a Tensor? (**rpimag.co/whatisatensor**)

> **FUN FACT!**
>
> The prevalence of matrices in neural networks and the use of matrix multiplication is why graphics cards are so useful for training neural networks. Matrices are conveniently useful for creating the 3D graphics that modern games rely on.

Python virtual environments

In the next section, you'll have a chance to get some hands-on coding experience with tensors. Before you can experiment with this code, you'll need to set up a Python virtual environment.

Python is installed on Raspberry Pi OS by default. To manage the third-party modules that you'll need in each chapter, you'll need to set up a *virtual environment* (a Python sandbox for installing modules without affecting your Python installation).

First, use the following commands to create a virtual environment in the **.virtualenvs** subdirectory under your home directory (we're calling it **tensorfun**, but you'll use this method to create separate environments for other projects):

1. Open a Terminal window and run this command:
   ```
   python3 -m venv ~/.virtualenvs/tensorfun
   ```

2. Activate the environment by running the following command:
   ```
   source ~/.virtualenvs/tensorfun/bin/activate
   ```

You *must* activate the environment each time you open a new Terminal window for it to take effect.

You can configure many code editors, such as Thonny and Visual Studio Code, to be aware of your virtual environment. This can help when you want to run a script from the editor but may also help if your editor checks the syntax for your code as you type:

Thonny

If you'd like to use Thonny to edit and run your Python code, click the Raspberry Pi menu, choose **Programming**, and click **Thonny**. Next, change the interpreter to your virtual environment. Click on **Local Python 3 • /usr/bin/python3** in the bottom-right and choose **Configure Interpreter**. This brings up the **Thonny options** dialogue. Next, click the '**…**' icon next to **Python Executable**.

When the file open dialogue box appears, select **Home**, then right-click in the list of files. Click **Show Hidden Files**, then navigate to the **.virtualenvs/tensorfun/bin** folder. Double-click the **activate** file there. Finally, Click **OK**.

Visual Studio Code

To use your environment with Visual Studio Code, make sure you've installed the Python extension from Microsoft (**rpimag.co/vscodepy**), and that you have a Python program open. If you've downloaded the example code from the book's GitHub repository (see *Welcome* on page v), you can open the **ch01** subdirectory and open one of the examples for this chapter.

If you'd rather type all the code yourself, press **CTRL+N** to create a new file, then press **CTRL+S** to save it. Now, give it a filename with the standard Python file extension (**.py**) and select **Python** from the filetypes menu in the **Save** dialogue, then save it in your working directory (see "A WORKING DIRECTORY OF YOUR OWN" on page 12). If you see a banner across the top indicating that Visual Studio Code is in Restricted Mode, click **Manage** and trust the current window.

Click your current **Python** version in the lower right of the Visual Studio window and you should see a popup menu as shown in Figure 1-1. It should contain all the virtual environments you've created. Pick the one named **tensorfun**. On Raspberry Pi OS, the Visual Studio Code Python extension automatically looks for virtual environments in the **.virtualenvs** folder in your home directory.

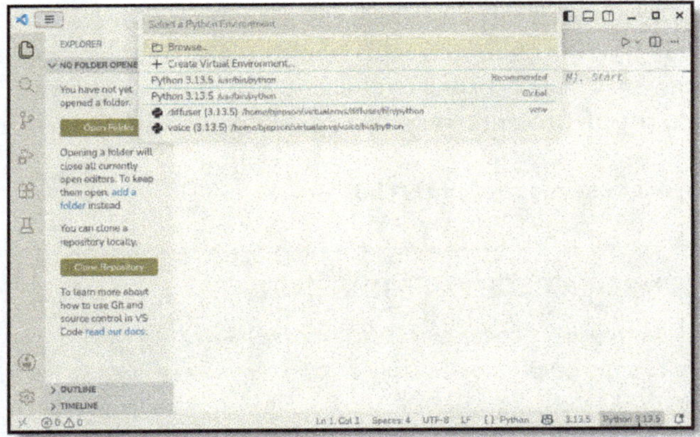

Figure 1-1   Choosing a virtual environment in Visual Studio Code

SCREENSHOT COLOURS

Both Visual Studio Code and the Raspberry Pi Terminal applications use a dark colour theme by default. While this is easier for some to read on-screen, it is a bit harder to read in a printed book or e-book. To improve readability, we've switched some of the screenshots in this book to the Solarized Light palette.

To change the Visual Studio code theme, press **CTRL+SHIFT+P** to open the Command Palette, type `Color Theme`, and choose **Preferences: Color Theme** from the popup menu, then pick a theme. You can change your palette in Terminal as well: click the **Edit** menu and choose **Preferences**, and make sure you're on the **Style** tab.

Experiment with TensorFlow in Raspberry Pi OS

A great way to understand tensors is through hands-on coding. You can find the code for this chapter in this book's GitHub repository (see *Welcome* on page v). Use `git clone` to check it out, then use the `cd` command to change to the **ai-projects-raspberrypi/ch01** subdirectory, as in:

```
cd ~/ai-projects-raspberrypi/ch01
```

A WORKING DIRECTORY OF YOUR OWN

If you want to type in the code yourself, use `mkdir` to create a working directory for the project, and `cd` to it (for example, `mkdir -p ~/aipi/ch01` followed by `cd ~/aipi/ch01`). Be warned, however: as you get further along in the book, the examples will get longer and more complex!

If you haven't already, you need to activate the virtual environment you created at the beginning of this section. After you activate it, add the Tensorflow and pillow packages (pillow is the Python Image Library, which we'll use to display an image):

```
source ~/.virtualenvs/tensorfun/bin/activate
pip install --no-cache-dir tensorflow pillow
```

If you're using the Terminal to run commands, type `python`, and run the following code at the prompt (if you're using Thonny, you can run these commands in the Shell area):

```
import tensorflow as tf
print(tf.__version__)
```

You should see the TensorFlow version number, such as "2.21.0". If you're running Python in the Terminal, type `exit` and press **ENTER** to return to the Terminal prompt (this is not needed for Thonny).

We can now write a Python program to explore the different tensor shapes. You can find the following example (**tensor_tf.py**) in the **ch01** directory:

```python
import tensorflow as tf
from PIL import Image

# Create a 0-dimensional (or rank 0) tensor
scalar = tf.constant(42)
print(scalar.shape) # Output: ()

# Create a 1D tensor (vector)
vector = tf.constant([1, 2, 3])
print(vector.shape) # Output: (3,)

# Create a 2D tensor (matrix)
matrix = tf.constant([[1, 2, 3], [4, 5, 6]])
print(matrix.shape) # Output: (2, 3)

# Create a 3D tensor (e.g., image with 3 colour channels)
image_tensor = tf.random.uniform(shape=[128, 128, 3])
print(image_tensor.shape) # Output: (128, 128, 3)
image_array = tf.keras.utils.array_to_img(image_tensor)
print(image_array.size) # Output: (128, 128)
image_array.show()
```

This program will first create a zero-dimensional tensor that can represent a single scalar value, then moves up on in dimensionality until we reach three dimensions. For the zero-, one-, and two-dimensional tensors, we're not doing anything with the values. But seeing as we're in three dimensions, we may as well create an image with our data. Higher dimensional tensors are possible, as well.

In the case of the three-dimensional **image_tensor**, we fill it with values from a TensorFlow function that generates random numbers between 0.0 and 1.0. The tensor's first two dimensions represent the coordinates of a two-dimensional image that's 128 by 128 pixels. The third dimension is three elements deep, and each element in that dimension represents one colour channel (red, green, and blue). We then use the **array_to_img()** function to turn it into an image. After that transformation, it becomes a two-dimensional array, where each element represents an RGB value.

If you're running this from the Raspberry Pi desktop, it should open the image, which is made up of random noise, in the Eye of MATE viewer as shown in **Figure 1-2**.

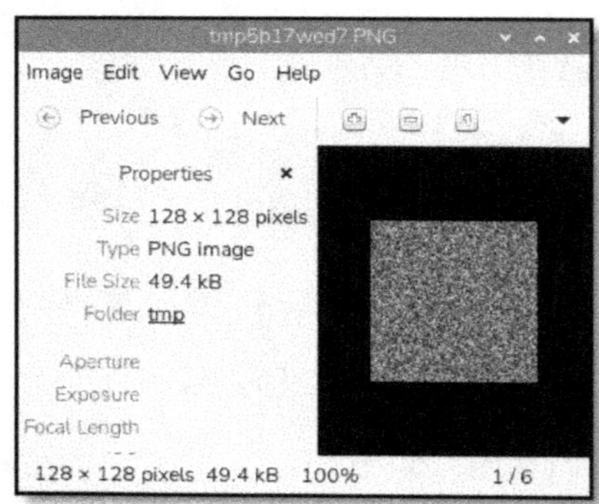

Figure 1-2   Viewing the randomly generated pixels

As you learn how to build models in later chapters, you won't be looking at tensors directly. The models are too complex, and the connections between tensors too numerous, for us to be able to make much sense of them. Nevertheless, it's helpful to understand what a tensor is, if only because it shows how little we have to fear from them.

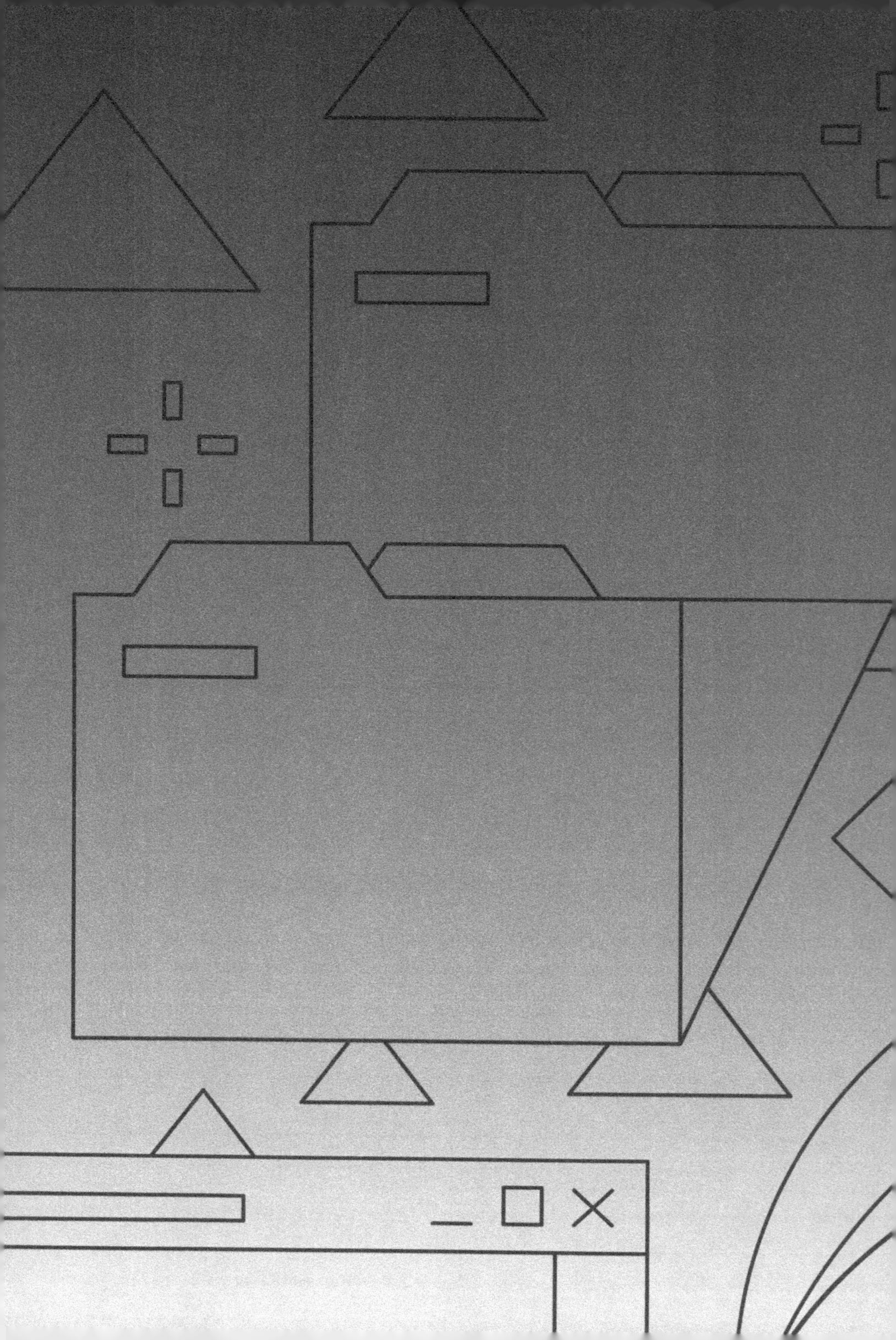

Chapter 2

The Raspberry Pi AI stack

The software and hardware you'll need to build your own
AI-enabled projects

It is worth mentioning that on their own, recent models of Raspberry Pi are capable AI devices. Even without a dedicated neural processing unit (NPU), there are many AI tasks that run perfectly well on a Raspberry Pi 4 or 5.

These native tasks include image recognition (either from image files or using a Raspberry Pi Camera Module), voice recognition, and generative large language models (LLMs) such as Llama and Deepseek. There are many projects from voice assistants to smart security cameras that don't require any form of hardware acceleration.

However, these AI models often run slowly on a Raspberry Pi without acceleration, and may not provide the real-time performance your project deserves. They also take up a large amount of physical RAM, leaving little room for the rest of your project.

An accelerator can speed things up. Without an accelerator, you might be able to perform image recognition on a still image, or generate text (slowly!) with an LLM, but with the right piece of hardware, you can recognise objects in video streams, generate text (responsively!) with an LLM or even a vision-language model (VLM), and have your Raspberry Pi manage a project in the home or on the edge of an industrial environment.

Raspberry Pi offers three different AI accelerators, in two categories:

Raspberry Pi AI Camera

> Like the Raspberry Pi Camera Module 3, this connects directly to Raspberry Pi via the Camera Serial Interface (CSI). The Raspberry Pi AI Camera uses the Sony IMX500 imaging sensor to provide low-latency, high-performance AI capabilities to any camera application. Tight integration with Raspberry Pi's camera software stack allows users to deploy their own neural network models with minimal effort.

Raspberry Pi AI HAT+ / AI HAT+ 2

> The AI HAT+ modules are add-on boards for Raspberry Pi 5 that come with a built-in AI accelerator chip: the Hailo neural processing unit (NPU). The Hailo NPU allows Raspberry Pi 5 to run hardware-accelerated AI models locally, removing the need to send data to a remote cloud server for processing.

Raspberry Pi AI Camera Performance

The Raspberry Pi AI Camera works differently from traditional AI-based camera image processing.

The camera module contains a small Image Signal Processor (ISP) which turns the raw camera image data into an input tensor. The camera module sends this tensor directly into the AI accelerator within the camera, which produces output tensors that contain the inferencing results. The AI accelerator sends these tensors to the Raspberry Pi. There is no need for an external accelerator, nor for the Raspberry Pi to run neural network software on the CPU.

At startup, the IMX500 sensor module loads firmware to run a particular neural network model. During streaming, the IMX500 generates both an image stream and an inference stream. This inference stream holds the inputs and outputs of the neural network model, also known as input/output tensors.

The IMX500 module supports real-time image inferencing at the following resolutions and framerates (measured in fps, or *frames per second*):

- Full resolution: 4056×3040 10-bit 10 fps. This mode offers the sharpest images, but a lower frame rate and weaker performance in dark environments.

- 2×2 binned: 2028×1520 10-bit 30 fps. This lower-resolution mode improves frame rates and is more sensitive to light than full resolution, resulting in less image noise even in low-light situations.

AI HAT+ variations

The following table lists the main capability differences between AI HAT+ and AI HAT+ 2 (shown in Figure 2-1).

Feature	AI HAT+	AI HAT+ 2
Accelerator chip (Hailo NPU)	Hailo-8L (13 TOPS, INT8) or Hailo-8 (26 TOPS, INT8)	Hailo-10H (40 TOPS, INT4)
Memory (RAM)	Uses the memory on a Raspberry Pi 5	Has its own 8 GB onboard memory, supporting LLMs and VLMs up to ~6 billion parameters
Large Language Models (LLMs)	Not supported	Supported
Vision-Language Models (VLMs)	Not supported	Supported
Use cases	Object detection, camera post-processing, robotics, moderate neural workloads	Everything available on AI HAT+ plus generative AI workloads, including local LLMs and VLMs

Figure 2-1   Raspberry Pi AI HAT+ 2

AI HAT+ Performance

AI HAT+ is available in the following tera-operations per second (TOPS, described in "Reach the TOP" on page 6) inference performance variants:

- 13 TOPS, built around the Hailo-8L neural network accelerator. This suits moderate workloads, with performance similar to that of the AI Kit.

- 26 TOPS, built around the Hailo-8 neural network accelerator. This supports larger networks, higher throughput, and running parallel AI models.

AI HAT+ 2 delivers 40 TOPS inference performance, built around the Hailo-10H neural network inference accelerator. This supports all AI HAT+ workloads and adds local LLM and VLM capabilities, enabling applications such as local chat over documents.

Choosing your Hardware stack

It's time to start building our Raspberry Pi 5 AI hardware stack. Although you can experiment with AI on any model of Raspberry Pi, we recommend a Raspberry Pi 5. Here's a recommended stack:

- **Raspberry Pi 5 4 GB/8 GB** — We've gone for the 8 GB model to have enough space to run larger models in RAM. AI HAT+ 2 will load models in its own 8 GB RAM, leaving your Raspberry Pi free to manage other aspects of a project. If you are using an AI HAT+ 2 for your LLM, a 4 GB Raspberry Pi 5 will suffice.

- **Active Cooler** — Keeping your Raspberry Pi cool is important when running inference models as overheating will cause Raspberry Pi OS to throttle performance. We're going for the official cooling option: **rpimag.co/activecooler**

- **32GB microSD card** — AI models are large by name, and large by nature. You will need at least a 32 GB microSD card to install Raspberry Pi OS and the models (**rpimag.co/sdcards**). Consider 64 GB or 128 GB options for larger projects with images and video files. You can also use Raspberry Pi's Flash Drive 128 GB/256 GB if you want extra storage: **rpimag.co/flash**

- **Camera** — We've connected a Raspberry Pi AI Camera to our build so we can run IMX500 object detection models inside the camera hardware (**rpimag.co/aicam**). You could use a standard Raspberry Pi Camera Module 3 (**rpimag.co/cam3**) with Ultralytics YOLO26 models on the CPU, or with an AI HAT+ or AI HAT+ 2 for a performance boost. Raspberry Pi Global Shutter Camera is optimal for industrial settings because it can capture rapid motion without introducing artifacts: **rpimag.co/gscam**

- **Tripod** — Our camera is placed on top of a Tripod Mount from The Pi Hut, connected to a Small Tripod via a Heavy Duty Tripod Swivel-ball adapter. See **rpimag.co/tripodhut**.

- **Case** — Our build is placed inside a Raspberry Pi Official Case (**rpimag.co/case**). You could also use a Bumper to keep things safe on your desk (**rpimag.co/bumper**).

M.2 HAT+ OR AI HAT+

You may be tempted to use Raspberry Pi's M.2 HAT+ solution to attach NVMe storage to your AI stack. While this is a fast, and spacious, storage solution it uses the same PCIe connection as the AI HAT+ and AI HAT+ 2. So you can't connect both at the same time. This is why we suggest a large microSD card or Flash Drive as the storage option.

Attaching the Raspberry Pi AI Camera

While the AI Camera (Figure 2-2) doesn't massively increase the amount of TOPS a Raspberry Pi has, its IMX500 sensor has 8 MB RAM on board to perform inferencing on the camera module. This avoids the need to send the image to the Raspberry Pi for processing. It instead sends the inferencing results via the camera cable to Raspberry Pi, keeping your Raspberry Pi free for other tasks. You'll learn how to use the AI Camera in Chapter 8, *Recognise video and image content*.

Figure 2-2   Raspberry Pi AI Camera

The Raspberry Pi AI Camera uses the Sony IMX500 imaging sensor to provide low-latency, high-performance AI capabilities to a variety of camera applications. Its tight integration with Raspberry Pi's camera software stack lets you deploy your own neural network models with minimal effort.

Raspberry Pi AI Camera attaches to the CSI port using an appropriate camera cable:

▸ Raspberry Pi flagship models up to and including Raspberry Pi 4 use the standard, 15-pin connector. For these boards, use the Standard-Standard camera cable provided with your camera.

▸ Raspberry Pi 5, all Raspberry Pi Zero models, and Compute Module IO boards use the mini, 22-pin connector. For these boards, use the Standard-Mini camera cable.

With the camera installed, ensure that your Raspberry Pi runs the latest software. Run the following command to update:

```
sudo apt update && sudo apt full-upgrade
```

The Raspberry Pi AI camera needs to download runtime firmware onto the IMX500 sensor during startup, so you must install the IMX500 firmware (skip this if you are using a Raspberry Pi Camera Module, High Quality Camera or Global Shutter Camera):

```
sudo apt install imx500-all
```

Now that you've installed the prerequisites, restart your Raspberry Pi:

```
sudo reboot
```

Attaching Raspberry Pi AI HAT+ / AI HAT+ 2

When an AI HAT+ or AI HAT+ 2 is connected to a Raspberry Pi 5, Raspberry
Pi OS automatically detects the on-board accelerator and offloads supported AI workloads to it. This includes using various camera and vision
frameworks to run vision-based neural network models on the NPU.

Set up Docker

Docker enables us to download images with containerised applications.
Containers are like a virtual machine, but they contain all the resources
needed to run a particular application and share the host kernel. Docker
makes it easy to download and run a sandboxed environment with all the
software components you need for a particular project. You'll use Docker in both Chapter 5, *Run a local LLM* and Chapter 8, *Recognise video and
image content*.

Docker is used by several complex applications in this book, including the web interface for LLMs using AI HAT+ 2 and Ultralytics YOLO26 models. So it is worth setting it up and becoming familiar with its commands.

Add Docker to apt

First make sure you have uninstalled any old Docker packages. Open a terminal window and enter:

```
sudo apt remove $(dpkg --get-selections docker.io \
    docker-compose docker-doc podman-docker \
    containerd runc | cut -f1)
```

CONTINUATION CHARACTER

The \ in a command line is known as the continuation character. Type \ followed by **ENTER** and your prompt will show > to indicate it's waiting for the next line. Or you can type the whole command out on one line by removing all \ characters. Alternatively, copy and paste the block of text.

Unless you have Docker already installed, **apt** will report that these packages are not found.

Next we add Docker's official GPG (GNU Privacy Guard) key to the **keyrings** folder. Before doing that, update **apt**, then make sure **curl** and **ca-certificates** and are installed (they should be installed by default):

```
sudo apt update
sudo apt install curl ca-certificates
```

Next, we make sure our keyrings directory has the correct permissions with 0755. This enables the file owner (our admin account) to read, write, and execute. Meanwhile, just read and execute permission is set for groups and others. We do this with an install command that is normally used for copying files but in this instance is being used to adjust permissions.

```
sudo install -m 0755 -d /etc/apt/keyrings
```

Now we use **curl** to download Docker's GPG key and place it into our **keyrings** directory with the filename **docker.asc**.

```
sudo curl -fsSL \
  https://download.docker.com/linux/debian/gpg \
  -o /etc/apt/keyrings/docker.asc
```

> **? CONFIRM THE KEYRING**
>
> Check in on the **docker.asc** file with `ls /etc/apt/keyrings`. You can also view the public key with `cat /etc/apt/keyrings/docker.asc`.

We now need to ensure that all users can read the `docker.asc` file. We do this with the more standard **chmod** command with **a+r** as options:

```
sudo chmod a+r /etc/apt/keyrings/docker.asc
```

Next comes a funky multi-line piece of code that creates a file called **docker.sources** in our /etc/apt/ directory and contains the details of the Docker repository. Enter the first line and you will see a **>** in terminal. Enter each line carefully and press **RETURN** after each one. Each line is added to the **docker.sources** text file until you enter **EOF** (at which point you return to the command line).

```
sudo tee /etc/apt/sources.list.d/docker.sources <<EOF
Types: deb
URIs: https://download.docker.com/linux/debian
Suites: $(. /etc/os-release && echo "$VERSION_CODENAME")
Components: stable
Signed-By: /etc/apt/keyrings/docker.asc
EOF
```

Finally, run **apt** to update the system and confirm that it is successfully connecting to the Docker repository:

```
sudo apt update
```

The **apt** output should include a line like this:

```
Get:5 https://download.docker.com/linux/debian trixie
  InRelease [32.5 kB]
```

Check docker.sources

It's easy to make a typing error when creating the **docker.sources** file with the **tee** command, so let's check its contents. Because the **Suites: line** grabs the **$VERSION_CODENAME** line from the **os-release** command, you should confirm that it indeed reads **trixie**. Run this command:

```
cat /etc/apt/sources.list.d/docker.sources
```

It should look like this:

```
Types: deb
URIs: https://download.docker.com/linux/debian
Suites: trixie
Components: stable
Signed-By: /etc/apt/keyrings/docker.asc
```

If there's a problem use vim or nano to edit the **docker.sources** file:

```
sudo nano /etc/apt/sources.list.d/docker.sources
```

Install docker

Now that the docker repository is in apt it's time to install the various elements. Enter this line in terminal:

```
sudo apt install docker-ce docker-ce-cli containerd.io \
    docker-buildx-plugin docker-compose-plugin
```

Docker should run automatically after installation. To verify that Docker is running, use:

```
sudo systemctl status docker
```

Press q to exit `systemctl` and return to the command line. Some systems may require you to manually start docker:

```
sudo systemctl start docker
```

Finally, verify that the installation is successful by running the hello-world image:

```
sudo docker run hello-world
```

If this is the first run it will pull the `library/hello-world` container from the Docker Hub. You will see a message containing:

```
Hello from Docker!
This message shows that your installation appears to be
working correctly.
```

Add user to Docker

Adding a user to the docker group will prevent you from having to use `sudo` with each docker command. First, create a docker group (don't worry if it already exists):

```
sudo groupadd docker
```

Now, add your user to the docker group:

```
sudo usermod -aG docker $USER
```

Sign out and back in again so that your group membership is re-evaluated or run the following command to activate changes to the group:

```
newgrp docker
```

Finally test docker again, now without **sudo**:

```
docker run hello-world
```

You should see the same `Hello from Docker!` message as before.

Next steps

Now that you have your hardware setup, you can explore the Raspberry Pi documentation for your hardware (**rpimag.co/docs**). Be sure to bookmark the AI Camera documentation (**rpimag.co/aicamdoc**) and AI HAT+ documentation (**rpimag.co/aihatdocs**). In the next chapter, we'll look at the fundamentals of predictive modelling. You won't need Docker or any AI accelerator hardware in that chapter, but it won't be long before you need one or the other.

Chapter 3

An introduction to predictive models

Learn about two foundational Machine Learning techniques: linear and logistic regression

Before you get into the AI projects that are the heart of this book, this chapter introduces some fundamental machine learning techniques that will help you understand modelling and inference: *linear* and *logistic regression*. Both techniques model the relationship between data that you already know (one or more *independent variables*) and data that you want to predict (the *dependent variable*). To build these models, you need historical data for both sets of variables. Using that data, these techniques produce models that, given new values for the independent variables, attempt to predict the value of the dependent variable.

Different methods work best in different situations. A linear regression model is used when changes in one variable are associated with consistent changes in another. For example, if one variable increases, the other tends to increase (or decrease) in a steady way. If you plot these values on an X/Y graph, the points won't usually form a perfect straight line, but they will cluster around a line that represents the model's best estimate of the relationship between the variables.

Logistic regression is used in a different kind of situation: when the value you want to predict is not a number, but a *category*. Sometimes this means a binary outcome such as yes or no, true or false, or one or zero. It can also answer the question "what group does this belong to?" Instead of predicting a value directly, logistic regression predicts the likelihood of a particular outcome.

When building any model, it's helpful to inspect graphs of historical data to see whether a relationship between variables might exist. This becomes more challenging when you have multiple independent variables, since a simple X/Y plot is no longer possible. Fortunately, as you'll see later in this chapter, there are tools and techniques that can help you determine whether meaningful relationships are likely to exist.

Figure 3-1 shows a plot of some sample data, using x (the horizontal axis) for independent variables and y (the vertical axis) for dependent variables. This data lends itself well to linear regression. In logistic regression, the relationships fit an S-shaped, or *sigmoid curve*. This can make patterns harder to spot by eye than in linear regression. While you can use plots to help guide your intuition, such relationships are usually not as immediately obvious as straight-line trends.

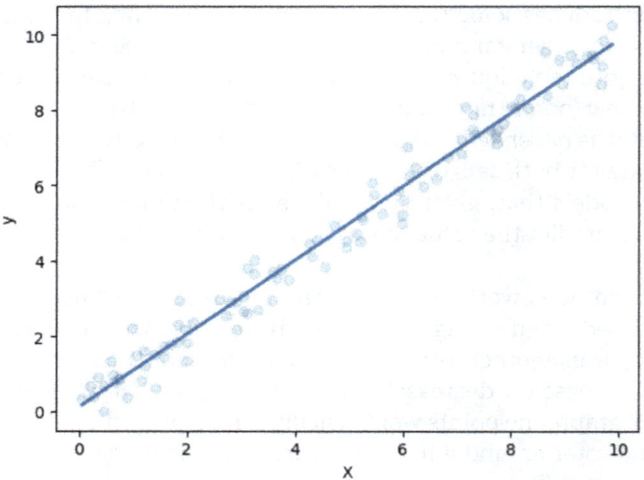

Figure 3-1   A plot that shows a linear relationship

Install Python modules

Before you can install the Python modules needed for this chapter, you'll need to create a virtual environment and activate it. For more details, see "Python virtual environments" on page 9:

```
python3 -m venv ~/.virtualenvs/Regression
source ~/.virtualenvs/Regression/bin/activate
```

After you've activated the environment, you'll need to install a few modules (you only need to do this once for each environment) with:

```
pip install jupyterlab pandas seaborn scikit-learn tqdm nbdime
```

The first of these modules provides the *Jupyter Lab* browser-based environment for creating interactive *notebooks*, which consists of *cells* that contain your code and documentation. The Jupyter Notebook web-based interface runs entirely on your Raspberry Pi.

The other modules you installed are used extensively with Jupyter:

pandas

> An open source framework for data analysis.

seaborn

> This is a module for data visualization. You'll use it to create charts and graphs.

scikit-learn

> This module offers a set of tools for predictive analysis, including support for linear and logistic regression.

tqdm

> This is a fast progress bar implementation for Python. It can draw progress bars at the Terminal or in your notebooks. Because you'll be running scripts that take a while to do their thing, `tqdm` can take the mystery out of how long a process will take to complete.

nbdime

> This module helps you compare changes in notebooks visually and can be used by the `git diff` command to compare changes you make against our original.

Collect some data

Before we can make some predictions, we need something to analyse. Your Raspberry Pi is always busy doing something and has some interesting metrics you can measure. Suppose you want to predict the temperature of your Raspberry Pi CPU — this will be your dependent variable. You'll need to pick some independent variables. But how do you know which ones? You can start by picking some likely candidates: CPU speed, CPU voltage, CPU utilisation (in percentage), and load average (the average number of processes using or waiting on the CPU).

You can use the `vcgencmd` command to measure the CPU speed, temperature, and voltage. The `psutil` Python module can measure load average and CPU percentage. If your virtual environment isn't active already, open a Terminal window on your Raspberry Pi, then activate it:

```
source ~/.virtualenvs/Regression/bin/activate
```

Next, install the `psutil` module in case it's not already installed:

```
pip install psutil
```

You can find this program in this book's GitHub repository (see *Welcome* on page v). Use `git clone` to check it out, then use the `cd` command to change to the **ai-projects-raspberrypi/ch03** subdirectory. If you'd rather create your own working directory and type in the code yourself, see "A WORKING DIRECTORY OF YOUR OWN" on page 12. The data collection script is named **collect_cpu_data.py**. First, it imports the modules needed to run this script:

```
import pandas as pd
import psutil
import os  # for os.popen to run vcgencmd commands
```

```python
import re # for regex to extract numbers from vcgencmd output
import time # for time.sleep to control sampling rate
from tqdm import trange # for progress bar
```

Next, you need to define some functions to get data out of **vcgencmd**. The
first one, **extract_number()**, uses a *regular expression* to extract numbers
from the output of **vcgencmd**.

In the following code, the regular expression **=(\d+(\.\d+)?)** first matches an equals sign (**=**). Then, the first item in the outer set of parentheses
matches one or more digits: **\d+**. The **\d** matches any digit, and the **+** after
the **\d** modifies the expression such that it matches one or more of the
preceding (one or more digits).

```python
# vcgencmd uses a format like "temp=45.2'C", so we need to
# extract the numbers with decimal points if any.
def extract_number(s):
    match = re.search(r"=(\d+(\.\d+)?)", s)
    return round(float(match.group(1)), 4)
```

The expression in the inner parentheses is similar, but it matches a decimal
point followed by one or more digits. The **\d** sequence matches any digit,
but because the full stop (**.**) is a special character in regular expressions
(it matches *any* character), we escape it by prefixing it with a ****.

The parentheses are used to combine matches (or parts of matches) into
numbered groups. By wrapping **\.\d+** in parentheses and adding the
question mark operator (**?**) after the closing parenthesis, it makes that entire expression optional, which allows the regex to match **=123** as well as
=1.23. By wrapping *everything* but the equals sign in parentheses, we can
refer to just the numeric part of the match with **match.group(1)**. In the

case of **volt=1.23**, group 0 would be the equals sign and everything after it, group 1 would be **1.23**, and group 2 would be the match from the inner set of parentheses (**.23**).

The remaining functions use **os.popen()** to run **vcgencmd**, then they call **extract_number()** to get the numeric result before returning it. The **cpu_speed()** function divides the result by one billion (10 to the 9th power, or **10**9**), before returning it. This converts the value (in Hz) to GHz.

```python
def cpu_speed():
    val = os.popen("vcgencmd measure_clock arm").read()
    return extract_number(val) / 10**9

def cpu_temp():
    val = os.popen("vcgencmd measure_temp").read()
    return extract_number(val)

def volts():
    val = os.popen("vcgencmd measure_volts").read()
    return extract_number(val)
```

Finally, we're ready for the main part of the program. The first line in this section checks to see this script was run directly (rather than being imported as a library). Next, it sets a progress bar (**pbar**) using **tqdm**'s **trange()** function. Then it runs through a loop for each value in **pbar**, and collects the data as an array, appending that array to the array of samples. Each time through, the script updates the progress bar with the most recent set of readings and then pauses for a quarter second.

```python
if __name__ == "__main__":
    numsamples = 2500
    samples = []
    pbar = trange(numsamples)

    for i in pbar:
        cpu_pct = psutil.cpu_percent()
        load_avg = round(psutil.getloadavg()[0], 2)
        r = [speed(), volts(), cpu_pct, load_avg, temp()]

        samples.append(r)
```

```python
    pbar.set_description(
        f"{r[0]:.2f}GHz {r[1]:.2f}V {r[2]}% {r[3]} {r[4]}C")
    time.sleep(0.25)

df = pd.DataFrame(samples)
df.columns = ['cpu_speed', 'volts', 'cpu_pct',
              'load_avg', 'temp']
df.to_csv(f"readings.csv", index=False)
```

After the loop, the script stores the data into a pandas **DataFrame** and saves it to a CSV file. Run this script with **python collect_cpu_data.py**, and use your Raspberry Pi normally while it's running: browse the web, watch some videos, and play some of the games from Code the Classics (you can install these from the **Recommended Software** utility). The idea here is to simulate what a normal user would do on a Raspberry Pi. Be sure to occasionally step away from the computer for a minute every now and then, so that you can include some baseline readings.

On a Raspberry Pi 3 or Zero, you will probably find that running this script consumes enough computing resources to greatly affect the readings. You may want to try adjusting the sleep time to a half a second (**0.5**) on one of those devices.

Each time you run the script, it will overwrite the **readings.csv** file. There is a sample file, **readings-sample.csv**, that contains results from running this on a Raspberry Pi 5.

Create your notebook

Now you're ready to create a Jupyter notebook. As with other examples in this book, you can download a ready-to-run notebook from the book's GitHub repository. However, we'll go through how to create the notebook step-by-step so you can learn how each part of it works.

Make sure you have a Terminal window open and that you've activated the **Regression** virtual environment (see "Python virtual environments" on page 9). If needed, **cd** to the same directory where you created the sample data file.

Run the command **jupyter lab**, and after a few seconds, your default web browser should open to the Jupyter Lab page, shown in **Figure 3-2**. You'll most likely see a few harmless error messages in the Terminal window while the notebook is running.

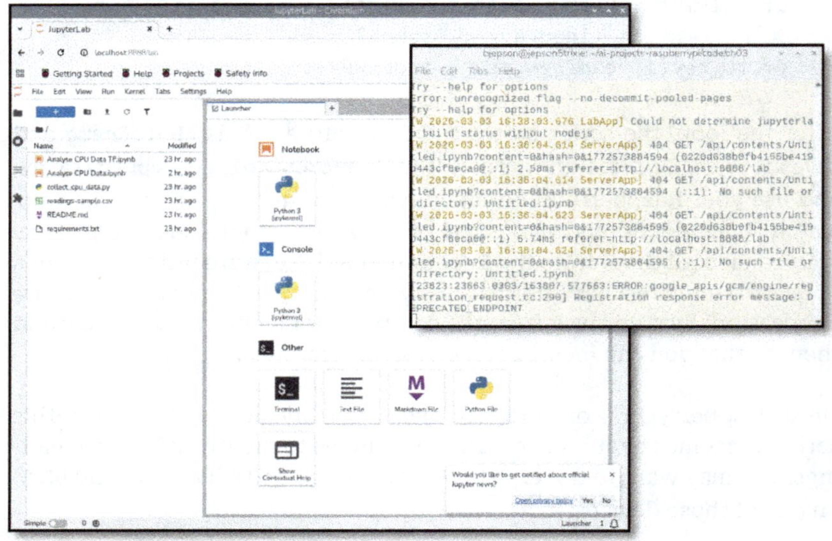

Figure 3-2   Jupyter Lab

The Jupyter Lab user interface is divided into two main panes. The left pane is a collection of vertical tabs, which defaults to the File Browser, which shows a list of files in your current directory. The right pane contains the Launcher tab as well as any notebooks that you have open. You can find the Jupyter menu just above those two panes. It includes several menus: **File**, **Edit**, **View**, **Run**, **Kernel**, **Tabs**, **Settings**, and **Help**.

To create a new notebook, you can click **Python 3 (ipykernel)** in the Notebook section of the Launcher. You'll be greeted by an empty notebook with a blank cell. You can type some Python code into this cell. Might we suggest **print("Hello, World!")**? You can run the code in the currently selected cell by pressing **SHIFT+ENTER**.

Jupyter Notebooks in Visual Studio Code

Although Jupyter usually runs in your web browser, Visual Studio Code can run Jupyter Notebooks. Unfortunately, the Thonny IDE *cannot* run Jupyter Notebooks. To run the notebooks in Visual Studio Code, make sure you've installed the Python extension and Jupyter extension from Microsoft (**rpimag.co/vscodepy** and **rpimag.co/jnvscode**), and that you have the Jupyter Notebook (a file ending in **.ipynb**) open.

If you've downloaded the example code from the book's GitHub repository (see *Welcome* on page v), you can open the **ch03** subdirectory to find the examples for this chapter. If you'd rather try typing in all the code, press **CTRL+SHIFT+P** to open the Command Palette, start typing **Jupyter**, and choose **Create: New Jupyter Notebook**.

Click the **Python** menu on the right side of the toolbar (just above the editor) and click **Select Another Kernel**. You should see a list of all the virtual environments you've created. Pick the one named **Regression**. On Raspberry Pi OS, the Visual Studio Code Python extension will look for virtual environments in the **.virtualenvs** folder in your home directory.

Linear regression

In linear regression, the dependent variable has *continuous values*, meaning it can take on any value within a range. For example, the temperature of a Raspberry Pi 5 CPU can range from about 0 °C to roughly 85 °C (you could push it higher, but the Pi would not be happy). When you obtained the current temperature using **vcgencmd measure_temp**, the result was some number within that range.

If you started Jupyter Lab in a local clone of this book's GitHub repository, you'll see one or more notebook files (with the **.ipynb** filename extension) in Jupyter's **File Browser**. You can open the first regression analysis notebook now: double-click **Analyse CPU Data.ipynb**. The notebook will open, as shown in **Figure 3-3**.

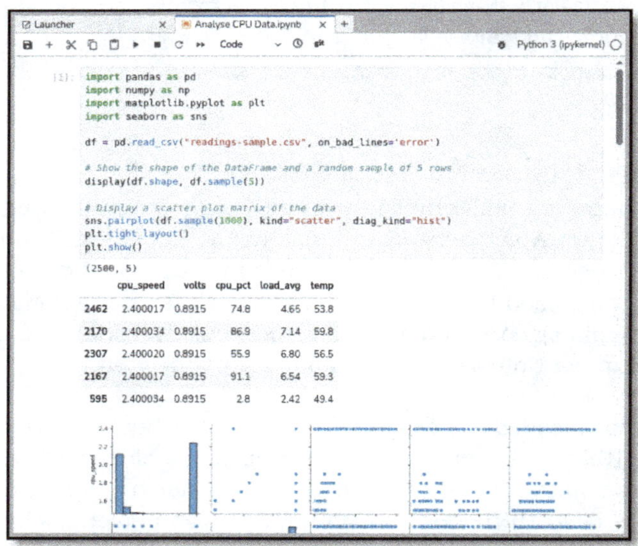

Figure 3-3   Opening the regression notebook

Load and inspect the data

The code in the first cell imports all the modules used in this notebook and then opens the sample CSV file (if you created your own CSV file in

"Collect some data" on page 34, change the filename if needed). It stores the contents of the file in a `DataFrame` object called `df`. Next, the notebook prints the shape of the data (2500 rows by 5 columns) and generates a *pair plot*, which is a table that shows the relationship between variables as well as the distribution of individual variables.

```python
import pandas as pd
import numpy as np
import matplotlib.pyplot as plt
import seaborn as sns

df = pd.read_csv("readings-sample.csv", on_bad_lines='error')

# Show the shape of the DataFrame and a random sample of 5 rows
display(df.shape, df.sample(5))

# Display a scatter plot matrix of the data
sns.pairplot(df.sample(1000), kind="scatter", diag_kind="hist")
plt.tight_layout()
plt.show()
```

Click in the cell to select it, then run it with **CTRL+ENTER**. You can see the complete pair plot in **Figure 3-4**. The distribution plots appear where the row and column labels refer to the same variable. Distributions start in the top left and continue diagonally down and across. The relationship plots appear in all the other intersection points. We're most interested in the last row, which shows the relationship between temperature and all the other variables. It should be easy to imagine a line that follows the plot of `load_avg` vs `temp`. With a little more effort, you can imagine such a line for `cpu_pct` (you don't have to imagine it, however — simply change the `kind` argument to `"regplot"` and it will draw a line).

However, for `cpu_speed` and `volts`, the connection between them and `temp` is not as clear. If you look at the distribution of both these variables (the upper-left plot and the intersection of the second column and row), you'll see that extreme high and low values make up most of the data.

While this makes it hard to ascertain the relationship, it doesn't mean there isn't one.

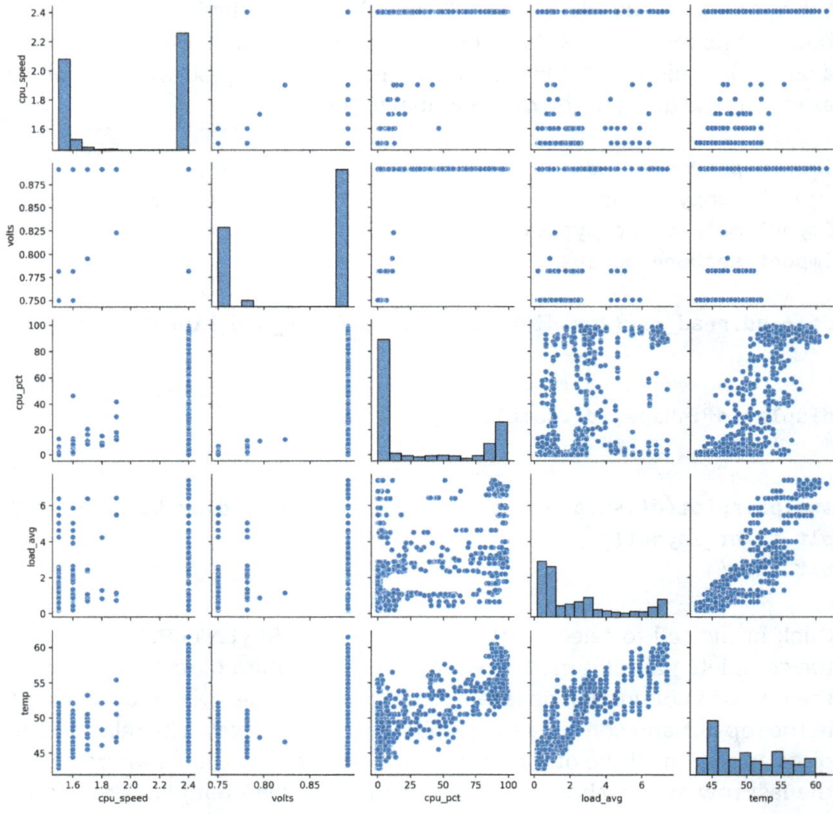

Figure 3-4   The pair plot from the notebook

You can use another type of visualisation, a heatmap, to display the value of a useful metric known as the *Pearson correlation coefficient*. This metric measures the linear correlation between two variables in the dataset, and ranges from 1 at the high end (perfectly correlated) to -1 (perfectly *negatively* correlated, meaning that as one value increases, the other decreases). The heatmap makes it easy to pick out high and low values.

Condition your data

Before you create the heatmap, it's worth *conditioning* the data a bit. For time series data like this dataset, a rolling average helps smooth out the data. This next cell applies a rolling average to produce a `DataFrame` named `df_adjusted`. It recalculates each value as a 5-point rolling average using the average of each value and the four values that precedes it. This means that the first four values in each variable can't be computed and are replaced with a missing data marker known as NA (not available). The code in this cell drops those values and then displays the heatmaps for both the original `DataFrame` (`df`) and `df_adjusted`. Be sure to run each cell by selecting it and pressing **CTRL+ENTER**.

```python
# Add a 5-point rolling average to smooth noise
df_adjusted = df.copy()
df_adjusted = df_adjusted.rolling(5).mean()

# Because of the rolling average, the first 4 rows will be NaN
# values. We can drop these rows.
df_adjusted = df_adjusted.dropna()
print(df_adjusted.shape)

# Display correlation for the original and adjusted DataFrames.
sns.heatmap(df.corr(), annot=True, cmap='BrBG')
plt.show()
sns.heatmap(df_adjusted.corr(), annot=True, cmap='BrBG')
plt.show()
```

The heatmap uses a colour palette known as BrBG, which uses shades of brown and green/blue to clearly highlight the differences in data. Each cell is also labelled with its correlation coefficient, as shown in **Figure 3-5**. You can see that the correlation improves noticeably with the rolling average applied. Notice that the values on the diagonal are always **1**, which makes sense: the values of a variable must be in perfect positive correlation with themselves.

But how should you interpret this correlation heatmap? There are two things to look for here. First, you want to choose independent variables (also called *features* in this kind of analysis) that correlate well to the dependent variable. Second, you may want to avoid using features that are tightly correlated to *each other*. In this heatmap, you can see that **volts** and **cpu_speed** are very well correlated, with a correlation coefficient of 0.95. It's not necessary to discard one of these features, but you need to keep an eye on them as you get further along.

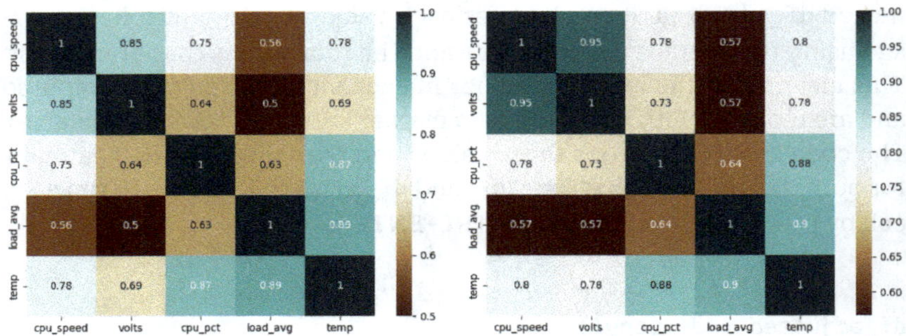

Figure 3-5 Correlation heatmaps for the original data (left) and for the data with a 5-point rolling average (right)

Perform the regression

With the data prepared, we're now ready to run the regression. The next cell in the notebook contains a single function named **do_regression()**. Before the function itself, it imports all the modules needed and then defines the function. Perhaps unsurprisingly, this is the function that will handle all the steps needed to actually run the linear regression.

The function's two arguments are a pandas **DataFrame** and a list of features. The first thing the code does is to separate the independent and dependent variables into **X** and **y**. **X** is a **DataFrame** containing just the features columns. The **y** variable is a pandas **Series**, which is essentially a single column from a **DataFrame**. It contains the values of the **temp** (temperature) column.

Next, the function uses the **train_test_split()** function to separate the data into training and test sets. You'll use the training data to train the model, and the test data to evaluate how well it's performing. **X_train** consists of about 75% of the rows from **X**, while **X_test** consists of the remaining rows. The same goes for **y_train** and **y_test**, but those values are taken from **y**. The code sets a **random_state** to an arbitrary number, which guarantees that **train_test_split()** returns the same results every time you run the code. This is useful if you want to experiment with the code a bit and ensure that you're working with the same training and test data set each time.

Next, the function creates a **LinearRegression** model and *fits* it to the training data. Fitting is the real work of the regression model: it determines a straight line that best fits the relationship between **X_train** and **y_train**. After that, it creates a set of predictions from the **X_test** data and stores it in **y_pred**. Because you separated the training and test data, you have a set of **y** values that are known to be correct, stored in the **y_test** variable.

The next step is to plot the predictions against the actual values and see how close the model came. The function creates a plot of actual vs predicted temperature and uses a scatter plot style with alpha blending of 20%. This means that each point in the plot will have 20% transparency, making it easier to spot where multiple points overlap because they'll appear darker than outliers.

```python
from sklearn.linear_model import LinearRegression
from sklearn.model_selection import train_test_split
from sklearn.metrics import mean_squared_error, r2_score
from IPython.display import HTML


def do_regression(df, features):

    # Separate the independent and dependent variables.
    X = df[features]
    y = df.temp

    # Split the data into training and test sets.
    X_train, X_test, y_train, y_test = \
        train_test_split(X, y, random_state=42)

    # Create and fit the model.
    model = LinearRegression()
    model.fit(X_train, y_train)
```

```python
# Plot the test data vs the model's predictions.
y_pred = model.predict(X_test)
sns.regplot(x=y_test, y=y_pred, scatter_kws={'alpha':0.2})
plt.xlabel('Actual temperature')
plt.ylabel('Predicted temperature')
plt.title(f"Feature(s): {', '.join(features)}")
plt.show()
```

The next cell specifies the features and calls **do_regression()** for each one. After it's done that, it makes one last call to **do_regression()**, this time with all the features. This last one is a multiple regression model that predicts a single dependent variable from multiple independent variables.

```python
# Specify the independent variables, also known as "features".
linreg_features = ["load_avg", "cpu_pct", "cpu_speed", "volts"]

# Build a single-variable regression model for each feature,
# then build a multiple-variable model for all the features.
for fc in linreg_features:
    do_regression(df_adjusted, [fc])
do_regression(df_adjusted, linreg_features)
```

Figure 3-6 shows the five plots resulting from this. The top four plots show the actual versus predicted temperature from the single-variable regression model. If the model's predictions match the actual values perfectly, the dots would line up perfectly with the diagonal line. Of the four, **load_avg** is the closest, but it's far from a straight line. The bottom plot shows the multiple regression model for all four features. This looks much better than any of the single-variable regressions.

Graphs and visual intuition alone are an insufficient basis for evaluating a model's performance. Fortunately, there are some helpful metrics we can use to evaluate a model. There are two useful metrics we can use to determine how well the model is performing: the *mean squared error* (MSE) and the *coefficient of determination* (also known as R^2).

MSE calculates the differences between actual and predicted values, squares them, and takes the average. The nice thing about squaring the differences is that it accentuates large errors, so even a few outlier values can increase the error score. The lower the MSE, the better. R^2 is a metric

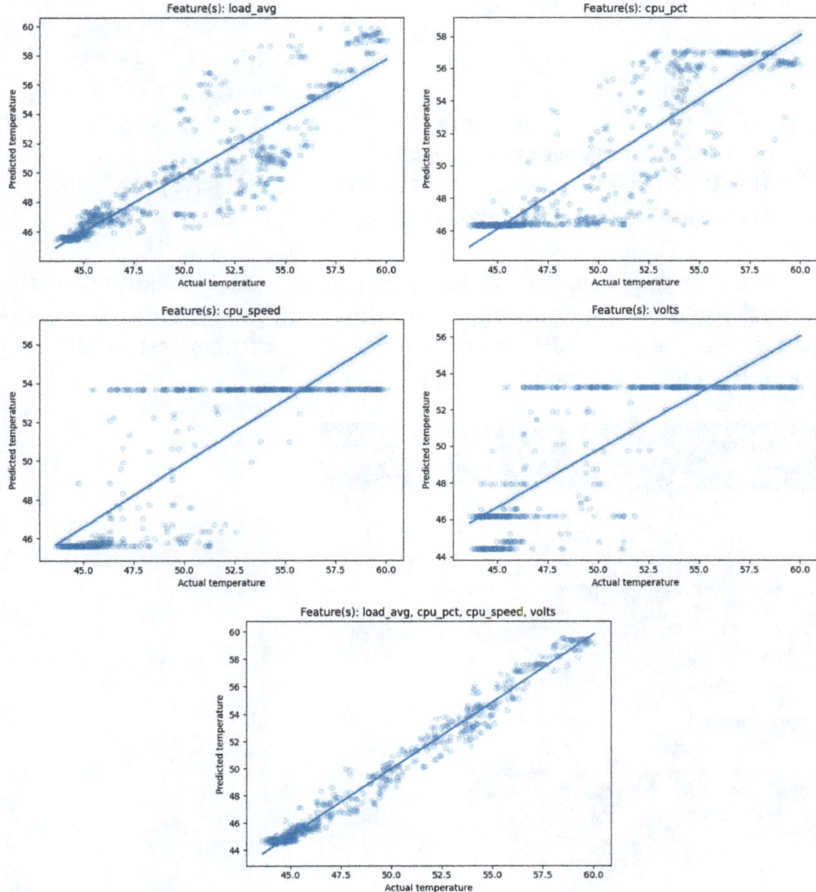

Figure 3-6   The five linear regression plots

that ranges from zero to one and is based on a formula that assesses how well the predictions and actual values compare. You want this as close to one as possible.

You can add the following code at the end of `do_regression()`. If you're editing this code on the fly, be sure to re-run the code in the cell to pick up the new definition of the function. (The sample code in the GitHub repository will already have this code in it.)

```python
# We want a low mean squared error and a coefficient of
# determination (R^2) as close to 1 as possible.
mse = mean_squared_error(y_test, y_pred)
r2 = r2_score(y_test, y_pred)
display(HTML(f"<b>Mean Squared Error:</b> {mse}"))
display(HTML(f"<b>R2 score:</b> {r2}"))
```

Run the cell that calls **do_regression()** again, and in addition to the graphs, you'll see the following output for each model. Here, it's abundantly clear that the multiple regression model performs best, confirming our earlier interpretation of the graphs.

Feature(s)	Mean Squared Error	R^2 score
load_avg	5.0797	0.7881
cpu_pct	5.3420	0.7772
cpu_speed	7.6249	0.6820
volts	8.4697	0.6467
All of them	0.4735	0.9803

Earlier, we said that when you fit a linear regression model, it determines a straight line that best fits the relationship between independent and dependent variables. If this sounds a lot like a mathematical linear function (**f(x) = ax + b**), it should, because it is! You can go back to the cell with the definition of **do_regression()** and add the following. Run it, and the following cell again, and you'll see some numbers you can play with (these lines of code are already in the example program in the GitHub repository):

```python
# Show some model stats
display(HTML(f"<b>Coefficients:</b> {model.coef_}"))
display(HTML(f"<b>Intercept:</b> {model.intercept_}"))

# Pick and test a few samples
test_val = X_test.sample(3)
prediction = model.predict(test_val)
display(HTML(f"<b>Test samples:</b>"))
display(HTML(test_val.to_html(index=False)))
display(HTML(f"<b>Predictions:</b> {prediction}"))
```

This new code displays two constants for the linear function: one or more coefficients (**a**); and the intercept (**b**). It also displays three randomly sampled values and the model's predictions. Here's the output from one of our test runs:

```
Coefficients: [1.99871962]
Intercept: 45.05711954089134
Test samples:
load_avg
    0.33
    1.28
    1.66
```

Let's use the linear function to calculate the predictions by hand. You can add an empty cell and paste this code into it, or better yet, work it out on paper, with a pocket calculator, or for extra points, a slide rule! Here, we're using some sample data for **load_avg**:

```python
a = 1.99871962
b = 45.05711954089134
for load_avg in [0.33, 1.28, 1.66]:
    print(load_avg * a + b)
```

The results are as follows, which matches the predictions!

```
45.71669701549134
47.61548065449134
48.37499411009134
```

How can we do this with multiple regression? We can use basically the same approach, but the equation is more complex. Here's some sample output for the multiple regression model:

```
Coefficients: [1.19572801 0.05043958 1.1171309 6.78998038]
Intercept: 37.480108349477504
Test samples:
load_avg.       cpu_pct      cpu_speed       volts
   6.836           87.44      2.400028      0.8915
   0.300.           0.54      1.500017      0.7500
   1.140.          90.14.     2.400026      0.8915
Predictions: [58.79895482 44.63426452 52.12427348]
```

Here's how you'd run that first prediction by hand:

```
b0 = 37.480108349477504 # Intercept
b1 = 1.19572801
b2 = 0.05043958
b3 = 1.1171309
b4 = 6.78998038
X = [6.836, 87.44, 2.400028, 0.8915]
print(b0 + b1*X[0] + b2*X[1] + b3*X[2] + b4*X[3])
```

This gives us 58.79895484947271. Given the vagaries of how floating-point numbers are stored and retrieved, this is as near as makes no difference.

Logistic regression

In a logistic regression analysis, the dependent variable has *categorical values*. For example, if you ask "Is the CPU voltage above 0.825 volts?", you get a no or a yes (a zero or a one). The answer to this question will never be 0.5, even though the model may estimate a 50% probability that the voltage is high. In practice, these probabilities are used to make a yes-or-no decision by choosing a cutoff point where a "no" turns into a "yes", such as 50%.

None of the variables in our dataset have a categorical value, but you can synthesize one easily. Suppose you want to indicate whether the processor voltage is high or low. You'll just need to define a threshold, perhaps 0.825 volts, after which you'd consider the voltage to be high. The next cell in the notebook imports some modules that you'll be using, defines a function to calculate the voltage level, and then creates a new DataFrame (**df2**) that replaces the **volts** column with a **voltage_level**. It then displays a correlation heatmap for the new dataset.

```python
from sklearn.pipeline import make_pipeline
from sklearn.preprocessing import StandardScaler
from sklearn.linear_model import LogisticRegression
from sklearn.metrics import confusion_matrix
from sklearn.metrics import ConfusionMatrixDisplay


def calculate_voltage_level(x):
    return 1 if x >= .825 else 0



df2 = df_adjusted.copy() # make a copy before changing anything

# Convert cpu voltage into a low/high indicator.
df2["voltage_level"] = df2.volts.apply(calculate_voltage_level)
df2.drop("volts", axis=1, inplace=True) # remove volts column

# Visualise correlation
sns.heatmap(df2.corr(), annot=True, cmap='BrBG')
plt.show()
```

As with the linear regression, we should look for features in **Figure 3-7** that correlate to the dependent variable (voltage level), but don't necessarily correlate too closely to one another. CPU speed, CPU percentage, and temperature correlate well to voltage level, but they are all somewhat well correlated to one another. Of these, temperature correlates the highest to both other features, so we can probably safely eliminate temperature. You can also turn to domain knowledge to construct an argument for eliminating temperature: in general, CPU speed is fully or in part controlled by the voltage level, so we wouldn't need a predictive model to determine one from the other.

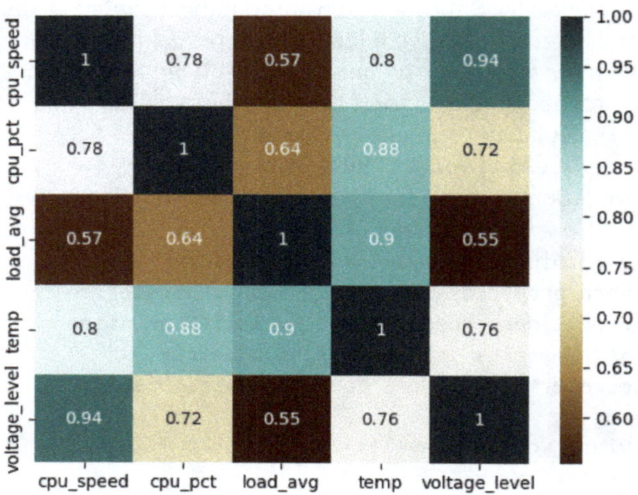

Figure 3-7   Correlation matrix for the new dataset

Here's the code for running a simple logistic regression between the selected features and dependent variable. As with the linear regression, the code isolates the X and y values and splits the data into training and test data:

```python
# Prepare the data
features = ["cpu_speed", "cpu_pct"]
X = df2[features].values
y = df2.voltage_level.values
X_train, X_test, y_train, y_test = \
    train_test_split(X, y, test_size=.2, random_state=42)

# Scale the data and run the regression.
logreg = make_pipeline(StandardScaler(),
                       LogisticRegression())
logreg.fit(X_train, y_train)

# Generate predictions based on the test inputs
y_pred = logreg.predict(X_test)
```

```python
# Evaluate the model with a confusion matrix
cm = confusion_matrix(y_test, y_pred)
disp = ConfusionMatrixDisplay(confusion_matrix=cm,
                              display_labels=logreg.classes_)
disp.plot()
plt.show()
```

Unlike the linear regression, this code uses the **make_pipeline()** function to create a two-part workflow: first, the data is scaled using scikit-learn's standard scaler, which transforms the data such that it has a mean (average) of 0, and a standard deviation of 1. So, for example, if we scaled the values 0, 10, and 100 with **StandardScaler**, we'd end up with -0.82, -0.59, and 1.41. These three numbers average out to zero, and if you calculate the standard deviation, you'll get 1.

STANDARD SCALER, STANDARD DEVIATION

You don't *need* to calculate the standard deviation, but if you wanted to, you'd subtract the mean from each value, square the result, add them all up and divide by the number of items in the list. Since the mean is zero, that's one less step:

$-0.82^2 + -0.59^2 + 1.41^2$

$0.67 + 0.35 + 1.99 \approx 3$

Of course, three divided by three gives you one, showing how **StandardScaler** makes good on its statistical promise while maintaining the proportional distance between data points. In both the original and the scaled data set, the distance between the second point and the first point is 10% of the distance between the last and first points.

The second stage of the pipeline runs the logistic regression. The call to the **fit()** function runs both pipelines in order. Once that's done, the model is complete, and the code creates a set of predictions based on the test inputs (**X_test**) and stores them in **y_pred**. You can evaluate the model by comparing the predictions to the correct values in **y_test**.

When working with logistic regression, you'll use different evaluation methods than what you used for linear regression. One of the most intuitive options is the *confusion matrix* (you may also want to look into scikit-learn's *classification report*).

You can see the confusion matrix in **Figure 3-8**. From upper left, moving clockwise, the cells are true negatives, false positives, true positives, and false negatives.

- A true negative occurs when both **y_pred** and **y_test** agree that, for a given input from **X_test**, the result is 0. There are 228 of these in the confusion matrix.

- A false positive occurs when **y_pred** (incorrectly) predicts a result of 1 where **y_test** has a 0. There's only one false positive.

- You get a true positive when **y_pred** and **y_test** agree, but in this case, they agree on a result of one. There are 263 of these.

- A false negative, of which there are 8, is when the model incorrectly predicts a result of 0, but the real value (**y_test**) is 1.

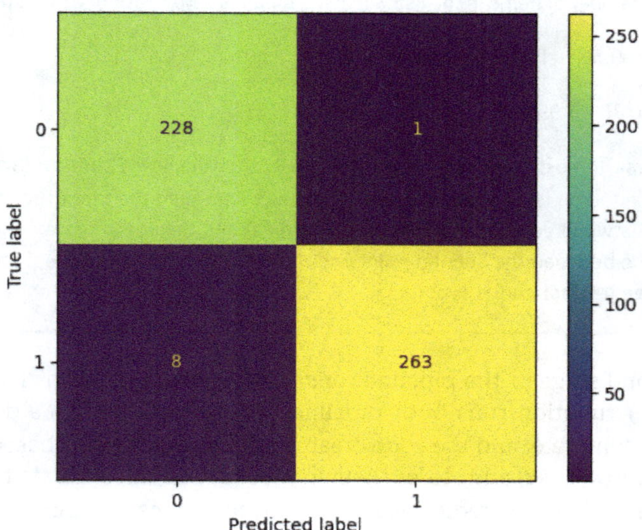

Figure 3-8　The confusion matrix

You can tell immediately that the model is doing quite well at predicting 0 (low voltage threshold) but is not quite as good at calculating when your Raspberry Pi has crossed the high voltage threshold (1). Another important detail is that the total number of low and high values is relatively well-balanced. There are 229 low values in the test data (total count of true negatives and false positives) and 271 high values (total count of true positives and false negatives). It's reasonable to conclude that the model performs well.

From regression to neural networks

A linear or logistic regression is essentially equivalent to a single-layer neural network. It takes inputs (features, or independent variables), applies a model to those inputs, and returns predictions. As with a logistic regression, a neural network layer can use an *activation function* to convert continuous probabilities into a 0 or 1.

In machine learning, neural network layers are stacked on top of one another with many hidden layers in between the input layer and output layer. Further, in neural networks, you can have different kinds of activation functions, and combinations of input, hidden, and output layers allow neural networks to generate more complex models to help make better predictions.

Although linear and logistic regression can't model the complex situations as neural networks, they are easier to interpret and adjust. As with every tool, your choice of linear regression, logistic regression, or neural network will depend on what problem you are trying to solve.

Chapter 4

Generate images from prompts

Image generation shows off the capabilities of Raspberry Pi in a colourful way

Image generation is one of the more fascinating (and controversial) applications of AI. Models such as Stable Diffusion are trained on several billion images. These *diffusion models* (also known as *diffusers*) are trained by adding noise to images with Gaussian noise. They then run the process in reverse, gradually predicting and removing noise from an image in iterations known as *steps*. This iterative denoising process produces an image that is coherent to humans.

In this chapter we've chosen to experiment with Stable Diffusion by Stability AI (**stability.ai/stable-image**). Stable Diffusion's code and model weights are open source, and it is released under a community license. Its image base is provided by LAION (Large-scale Artificial Intelligence Open Network) who have an image base of five billion images (two billion were used to train Stable Diffusion). You can view more information at **laion.ai**.

Mitsua Diffusion One (**rpimag.co/mitsua**) is a Stable Diffusion-compatible text-to-image model trained exclusively on public domain content. This was trained on a much smaller dataset of a 100 million images. Researchers have found that some models, including Stable Diffusion, did not ask consent of artists prior to training, which remains a contentious issue. Using Mitsua Diffusion One offers a more ethical choice.

We will use Hugging Face to download the models, and Hugging Face's diffusers library to run the diffusion models. Running the full Stable Diffusion model requires a Raspberry Pi 4 or 5 with 8 GB RAM (16 GB is preferable) and at least 64 GB of storage (models are large though, so 128 GB or 256 GB microSD cards will enable you to experiment with more than one model at a time). You'll also need an internet connection to download the models (but after you download them, you can run them offline).

First, however, we will look at a library called OnnxStream by Vito Plantamura (**rpimag.co/onnxstream**). This decouples the inference engine from the weights. Models like Stable Diffusion contain over a billion weights; which the inference engine uses to determine what each pixel should look like at each step. Typically, these are all stored in RAM alongside the inference engine. With OnnxStream, just the inference engine is stored in RAM, and the weights are streamed into RAM as they are needed.

This enables a version of Stable Diffusion to run on a Raspberry Pi Zero 2 with just 512 MB RAM or a Raspberry Pi 5 with 1 GB RAM. On a Raspberry Pi with 8 GB it leaves you with enough space to run concurrent compute tasks on the CPU.

Image generation lends itself perfectly to projects that generate a prompt and display an image on a Raspberry Pi Display. You could grab weather data from an API and use it to generate the day's weather. Or a to do list that shows your next task visually. We've also seen an interesting project that uses a Raspberry Pi camera without a lens that grabs data from its current location, surrounding buildings, the time and weather, and then "guesses" what the view is like.

Let's get started with image generation. You'll need:

- ▸ Raspberry Pi Zero 2, or Raspberry Pi 3B/4/5 for OnnxStream.

- ▸ Raspberry Pi 5 with 8 GB RAM for Stable Diffusion and other models (16 GB to run at full speed).

- ▸ 64 GB storage for OnnxStream/128 GB or more for Stable Diffusion.

- ▸ Hugging Face account (**huggingface.co**).

- ▸ Stable Diffusion 1.5 (**rpimag.co/sdv1-5**) or Mitsua Diffusion One (**rpimag.co/mitsua**). We'll show you how to download these later.

- ▸ Internet connection.

Get Stable Diffusion model with OnnxStream

We're going to start with a version of Stable Diffusion called OnnxStream. This is a custom version of the diffusion model that separates the inference engine from the weights.

OnnxStream enables you to run Stable Diffusion v1.5 and even SDXL in small RAM spaces. You can even run it on a Raspberry Pi Zero 2 with just 512 MB RAM. The downside is that this increases the image generation time as image generation makes heavy use of our CPU resources.

It is an ideal test case for Raspberry Pi 5 with 1 GB RAM, however, which blends four fast CPU cores with enough RAM to hold our inference engine and stream our weights.

We can even use Stable Diffusion XL 1.0 (SDXL) in a limited memory space. SDXL is a much larger model that generates 1024×1024 resolution images on Raspberry Pi. In this limited space it is advised to boot into the Command Line Interface (CLI) rather than the full Graphical User Interface (GUI). This frees up around 350 MB of RAM which on a device with limited RAM, such as the Raspberry Pi Zero 2 with 512 MB RAM or Raspberry Pi 5 with 1 GB RAM is a substantial part. Running in headless mode will prevent our Raspberry Pi 5 with 1 GB from using too much of the swapfile space.

The downside to using OnnxStream versus loading the Stable Diffusion full model is the time it takes to generate images. This is around fifteen minutes on a Raspberry Pi with 8 GB RAM and up to 11 hours on a Raspberry Pi Zero 2 with 512 MB. All of this is a lot more than the three to five minutes you will experience running the full version of Stable Diffusion v1.5 on a Raspberry Pi 5. So, you trade the memory requirement for speed.

You can find OnnxStream on Vito Plantamura's GitHub page: **rpimag.co/ onnxstream**. It is a great way to play around with image generation in a memory constrained environment. It also comes packed with features and makes downloading and using Stable Diffusion V1-5 extremely easy.

Switch to headless mode (optional)

If you're using a Raspberry Pi 5 with 4 GB or less of RAM, you'll want to make as much memory available as possible. Start with a fresh installation of Raspberry Pi OS Lite or turn off the GUI. If you are in the GUI, open a terminal window and enter:

```
sudo raspi-config
```

Choose: **1 System Options** then **S5 Boot** and select **B1 Console Text Console**. Next, choose **Finish** and answer "Yes" to the "Would you like to reboot now?" question.

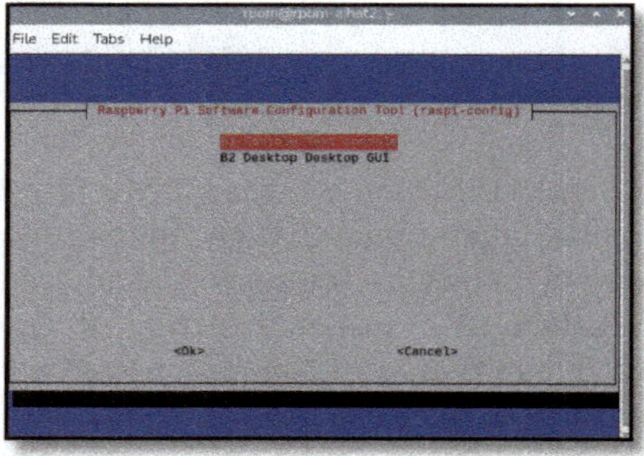

Figure 4-1 Using raspi-config to switch to headless mode

Build OnnxStream

We need to build the code from scratch so let's get the workspace set up and make sure you have all the requirements. We're going to put the files in a folder called **Programs**. Open a Terminal or SSH into your Raspberry Pi. Next, create and change into the folder:

```
mkdir ~/Programs
cd ~/Programs
```

Now make sure your system is up to date and has the software packages needed to compile OnnxStream:

```
sudo apt update
sudo apt full-upgrade
sudo apt install build-essential git cmake python3
```

Raspberry Pi OS should already have the latest versions of most of these. It will need to install `cmake` and its dependencies if you haven't already installed them. Now clone the GitHub repo and build the files from source:

```
git clone https://github.com/vitoplantamura/OnnxStream.git
cd OnnxStream/src
mkdir build
cd build
cmake ..
cmake --build . --config Release
```

The command to run OnnxStream's version of Stable Diffusion is `sd`. You will find it located in **~/Programs/OnnxStream/src/build**. After you finish running the preceding commands, your current directory should be this folder, but you can change back to it with:

```
cd ~/Programs/OnnxStream/src/build
```

You can now run `sd` by using the `./sd` command. First let's see what commands are available with:

```
./sd --help
```

This will list all the commands available. Here are the ones we are most interested in:

`--xl`

Runs Stable Diffusion XL 1.0 instead of Stable Diffusion v1.5.

`--turbo`

Runs Stable Diffusion Turbo 1.0 instead of Stable Diffusion v1.5.

`--models-path`

Sets the folder containing the Stable Diffusion models.

`--output`

Sets the output image file.

`--prompt`

The positive prompt. Specifies the image you'd like to generate.

`--neg-prompt`

Sets the negative prompt. Use this to specify attributes you'd like to avoid in the generated image.

`--steps`

Sets the number of diffusion steps. Default is 10.

`--ram`

Loads the entire UNET model into RAM for faster inference.

`--download`

A[uto] / **F**[orce] / **N**[ever] (re)download current model.

`--rpi`

A[utodetect] / **F**[orce] / **D**[isable] to configure the models to run on a Raspberry Pi.

`--rpi-lowmem`

Configures the models to run on a Raspberry Pi Zero 2.

Set sd instead of ./sd

We place `./` in front of **sd** to inform the OS to use the file in the current folder (indicated with a single `.`). This prevents the OS from heading off into the usual **PATH** environment variables to look for the **sd** command.

We can add an alias to the **sd** file to our .bashrc file so we can access **sd** from anywhere. Open the file for editing with **nano ~/.bashrc**.

Add this line underneath the **# Alias definitions.**:

```
alias sd='~/Programs/OnnxStream/src/build/sd'
```

Save the file with **CTRL+O** and exit **nano** with **CTRL+X**. Then, reload .bashrc with **source ~/.bashrc**.

Now it's time to try it out. First let's create a **Models** folder and change into it, then tell **sd** to download the weights for the default model, which is Stable Diffusion v1.5:

```
mkdir ~/Models
cd ~/Models
sd --download
```

When OnnxStream finishes the download it will automatically run the default prompt "a photo of an astronaut riding a horse on mars" with 10 steps. Each step will take around a minute to run on a Raspberry Pi 5 with 1 GB of RAM.

CHECK IN WITH HTOP

You can monitor **sd**'s resource utilisation with **htop** (**Figure 4-2**). Open another Terminal or SSH connection to your Raspberry Pi. If you are running in headless mode, you can switch to another console with **CTRL+ALT+F2** (you'll need to log in). Use **CTRL+ALT+F1** to return to your default console.

Next, run the command **htop** and press **SHIFT+M** to sort by memory usage. You'll see the **sd** command taking up all your CPU processing power and a chunk of your physical memory.

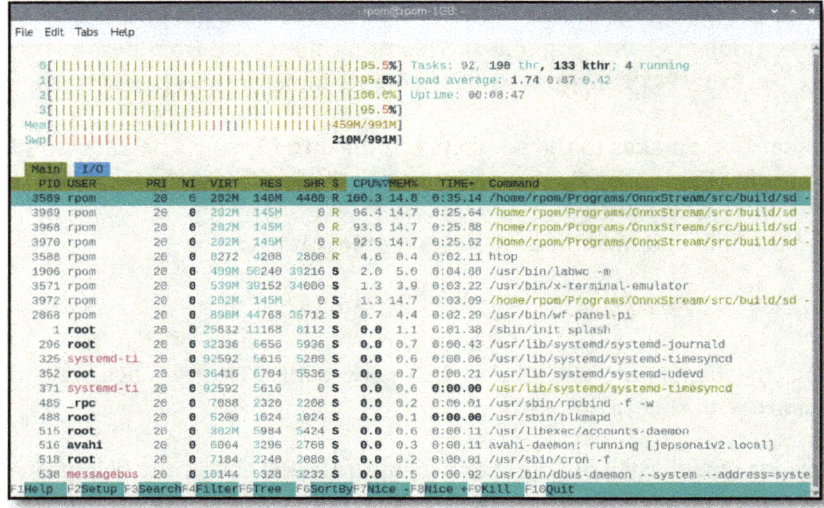

Figure 4-2 Checking CPU and RAM usage with htop

When it is finished, it will save the file as **result.png**. Because we are running in headless mode you can either swap back to the GUI to view it or use sftp from another computer running a GUI to grab the file.

Take a note of the **user@hostname** value in the command line (ours is **rpom@rpom-1GB)** and use this to grab the file from another computer:

```
sftp <user>@<hostname>
sftp> cd Models
sftp> get result.png
sftp> exit
```

Open the **result.png** file and take a look. You should have a pretty neat looking astronaut riding a horse as shown in **Figure 4-3**.

Figure 4-3 The default Stable Diffusion
OnnxStream prompt of an astronaut riding a horse

BLANK IMAGE?

Sometimes you will get a warning that reads: "Potential NSFW content was detected in one or more images." A black image will be returned instead (**Figure 4-4**). Try again with a different prompt.

Prompt your model

Now we have our first image we can look at the model. We can check its file size with the **du** command:

```
cd ~/Models
du -bsh *
```

This will show our **results.png** file and the **stable-diffusion-1.5-on-nxstream** directory, clocking in at 2.2 GB. This is a lot more space efficient than other models. Let's use our **sd** command to create more images.

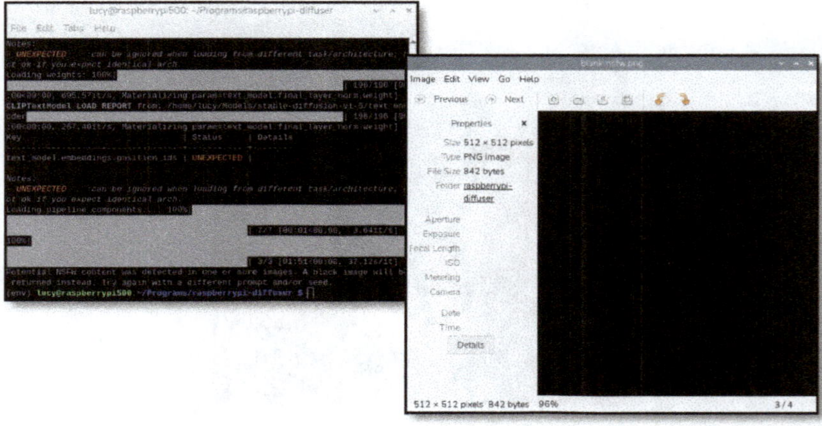

Figure 4-4 Images deemed NSFW (Not Safe For Work) appear as a blank PNG

Let's move to the home folder. Because we created an alias, we now can run **sd** from anywhere and use command line options to tell **sd** to look for models in the **Models** folder. We can also tell it to store our images in the **Pictures** folder.

```
sd --prompt "a photo of a black cat with green eyes" \
  --neg-prompt "blurry" --models-path ~/Models \
  --output ~/Pictures/black-cat.png --steps 3 --rpi-lowmem
```

This produces a picture of a black cat in our **Pictures** folder. It should take around three minutes. Let's break the commands down:

- **sd** runs the stable diffusion model

- **--prompt** what we want the image to contain

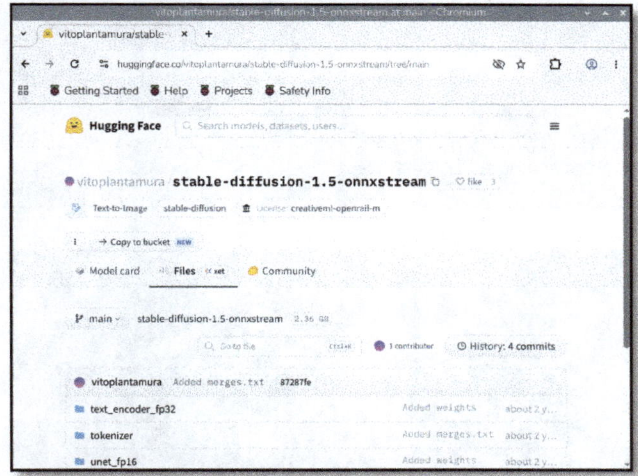

Figure 4-5   The Hugging Face website

- **--neg-prompt** what we want the image to avoid

- **--models-path** the location where we downloaded the **stable-diffusion-1.5-onnxstream** model

- **--output** where to save the image filename

- **--steps** how many steps the diffusion process should take

- **--rpi-lowmem** optimises the model for Raspberry Pi Zero 2 (which also works on our 1 GB Raspberry Pi 5).

Experiment with different prompts and steps. Around 30 to 40 steps generally produces good results. If you have a Raspberry Pi with 4 GB RAM or more, you can replace **--rpi-lowmem** with **--rpi**.

THE 35 STEPS

We find 35 steps produces good quality images and takes around 1 hour to produce an image on a Raspberry Pi 5 with 1 GB. **Figure 4-6** shows images of a "black cat with green eyes" generated with Stable Diffusion v1.5 using 5, 10, and 35 steps. This demonstrates how each step denoises the image further to remove randomness and add clarity.

Figure 4-6　Clockwise from top left: 5, 10, and 35 steps

Using Stable Diffusion XL

Our OnnxStream installation can also run Stable Diffusion XL. This is a much more recent and powerful model. It takes longer to run, but the images have a larger resolution (1024×1024) and tend to have a higher quality.

You download Stable Diffusion XL in OnnxStream using the `--xl` flag:

```
sd --download --models-path ~/Models --xl
```

The model will download the weights and then generate the default image of an astronaut riding a horse on mars with 10 steps. This took around
40 minutes on our Raspberry Pi (see **Figure 4-7**). It shows innate higher
quality than the regular SD model. Some elements are still fuzzy which
could be improved upon using more steps.

Figure 4-7   The horse-riding astronaut generated
using 10 steps with Stable Diffusion XL

Once OnnxStream has finished downloading and testing the model you
can run it on Raspberry Pi by adding the `--xl` flag to your command:

```
sd --xl --prompt "a photo of a black cat with green eyes" \
   --neg-prompt "blurry" --models-path ~/Models \
   --output ~/Pictures/black-cat-sdxl.png \
   --steps 35 --rpi-lowmem
```

Expect it to take around three hours to run completely. But the result
should be a pleasingly convincing cat image (**Figure 4-8**) generated on a
Raspberry Pi even with just 1 GB RAM.

Figure 4-8   An image of a black cat with green eyes generated by Stable Diffusion XL in 35 steps

Moving to full Stable Diffusion in RAM

If you want to cut down the time it takes to generate images, you'll need a Raspberry Pi 5 with at least 8 GB RAM. You'll also need at least a 128 GB SSD or microSD card because we will download the full model including the weights.

Install Hugging Face CLI

We're going to get our models from the popular Hugging Face website (**huggingface.co**). This website has become central to the open-source AI community. It contains a Model Hub where we can browse, and access information on models. There are also datasets for training and fine-tuning models, along with the popular Transformers library (a full-featured framework for working with models).

You can browse models on the Hugging Face website. We're going to be using Stable Diffusion v1.5 here (**rpimag.co/sdv1-5**).

There are a few different ways you can download models from Hugging Face. One way is with the Hugging Face Command Line Interface (CLI), which lets you easily download models and their weights. Hugging Face CLI is part of Hugging Face Hub (**rpimag.co/hfcli**). You can install it from the command line with:

```
curl -LsSf https://hf.co/cli/install.sh | bash
```

When it has finished, you will need to either close and open a new Terminal window or reload your **.bashrc** file to add the Hugging Face CLI to your **PATH** environmental variable:

```
source ~/.bashrc
```

Use the **hf download** command from Hugging Face CLI to download the Stable Diffusion v1.5 model to the **~/Models** folder. We pass in the user ID for the model: **stable-diffusion-v1-5/stable-diffusion-v1-5**, and the **--local-dir** option saves it to the ~/Models folder instead of a local cache. It's a good idea to run a check of the download first, so we also add the **--dry-run** option:

```
hf download stable-diffusion-v1-5/stable-diffusion-v1-5 \
  --local-dir ~/Models/stable-diffusion-v1-5 --dry-run
```

If everything goes well, repeat the command without **--dry-run**:

```
hf download stable-diffusion-v1-5/stable-diffusion-v1-5 \
  --local-dir ~/Models/stable-diffusion-v1-5
```

This will download 47.3 GB of files, so expect it to take a while. If you're using a Raspberry Pi 5 with 4 GB RAM, add the **--max-workers=1** argument to your hf download command. This will tell the Hugging Face CLI to only initiate one download at a time and will keep your Raspberry Pi from running out of memory.

> **? HUGGING FACE TOKENS**
>
> You can speed things up by signing up for a Hugging Face account and logging in to the command line with a token generated on your Settings page. See **rpimag.co/hf-tokens**.

```
rpom@rpom-aihat2:~ $ hf download stable-diffusion-v1-5/stable-diffusion-
v1-5 --local-dir ~/Models/stable-diffusion-v1-5 --dry-run
[dry-run] Fetching 36 files: 100%|████| 36/36 [00:00<00:00, 212.75it/s]
Download complete: : 0.00B [00:00, ?B/s]                [dry-run] Will dow
nload 17 files (out of 36) totalling 45.1G.
FILE                                                    SIZE
-----------------------------------------------------   ------
.gitattributes                                          -
README.md                                               -
feature_extractor/preprocessor_config.json              -
model_index.json                                        -
safety_checker/config.json                              -
safety_checker/model.fp16.safetensors                   -
safety_checker/model.safetensors                        1.2G
safety_checker/pytorch_model.bin                        1.2G
safety_checker/pytorch_model.fp16.bin                   -
scheduler/scheduler_config.json                         -
text_encoder/config.json                                -
text_encoder/model.fp16.safetensors                     -
text_encoder/model.safetensors                          492.3M
text_encoder/pytorch_model.bin                          -
text_encoder/pytorch_model.fp16.bin                     -
tokenizer/merges.txt                                    -
tokenizer/special_tokens_map.json                       -
tokenizer/tokenizer_config.json                         -
tokenizer/vocab.json                                    -
```

Figure 4-9 Downloading a model from Hugging Face using the hf download command

Install pip packages

Next, create the virtual environment:

```
python3 -m venv ~/.virtualenvs/diffuser
```

Activate the virtual environment so you can install modules with **pip**:

```
source ~/.virtualenvs/diffuser/bin/activate
```

We will now install the packages we need to run our Stable Diffusion model. There are four:

- Diffusers is a library for using pre-trained models to generate images and audio. It's versatile: other uses include upscaling the resolution of images, editing parts of an image, creating variations of images, and so on (**rpimag.co/diffusers**).

- PyTorch is an extensive tensor library for deep learning models (**rpimag.co/pytorch**).

- Transformers is a model-definition framework that supports machine learning with text, computer vision, audio, video, and multimodal models.

- Torchvision is an optional library that helps PyTorch work with video and audio. Some of the models you use may perform better if it is installed.

All can be installed with a single **pip install** command:

```
pip install --upgrade diffusers[torch] transformers torchvision
```

You can find the sample code from this chapter in this book's GitHub repository (see *Welcome* on page v). Use **git clone** to check it out, then use the **cd** command to change to the **ai-projects-raspberrypi/ch04** subdirectory, as in:

```
cd ~/ai-projects-raspberrypi/ch04
```

Create an image with Stable Diffusion v1.5

We now have our diffuser, transformer, and model and are ready to generate our first image. You can find the following example program in the **ch04** directory as **astro-horse.py**:

```python
import os # We need this to expand ~ to your actual home dir
from diffusers import StableDiffusionPipeline

# Get the expanded path to the model
model_name = "stable-diffusion-v1-5"
model = os.path.expanduser(f"~/Models/{model_name}")

# Load the model
pipe = StableDiffusionPipeline.from_pretrained(model,
                                    low_cpu_mem_usage=True)
pipe = pipe.to("cpu") # Move the model to the CPU

prompt = "an astronaut riding a horse" # Set the prompt
neg_prompt = "blurry"
image = pipe(prompt, negative_prompt=neg_prompt,
            num_inference_steps=5,
            width=640, height=640).images[0]

image.save("astro-horse.png")
```

Run it with **python astro-horse.py**. The model will take around three minutes to run depending on your setup. At the end you will find a file named **astro-horse.png** in your **ch04** folder. Open it to find out what image was generated (see **Figure 4-10** for an example).

HAVE A RASPBERRY PI 5 WITH 16 GB?

If you are using a Raspberry Pi with 16 GB RAM, you can set **low_cpu_mem_usage=False**. This will improve image generation time because it can load the model all at once. During each step, RAM usage falls to 2 GB, though you risk running into an Out-of-Memory (OOM) error if you do this with other models.

Figure 4-10 An image of an astronaut generated
using Stable Diffusion v1.5 on Raspberry Pi 5

You'll probably find the image of an astronaut riding a horse not to be particularly compelling. This first run had just five inference steps, and successive steps improve upon the quality of the image. Typically, you'll want to run between 30 and 40 steps. Open the code with an editor such as Thonny or Visual Studio code and change `num_inference_steps=5` to `35`. Now save the file and run it again with `python astro-horse.py`. You can have fun experimenting with the prompt and the number of inference steps to generate a variety of images.

Testing out other models

We have provided another program file in the GitHub repository called **astro-horse35.py**. This is set to run a specified model to create an astronaut riding a horse in 35 steps. The resulting file output will have the time it took to generate the image appended to the filename. Typically, Stable Diffusion v1.5 takes around fifteen minutes per image. You can download different models from Hugging Face and adjust the code to reference the new model.

For example, you can use Hugging Face to download the Mitsua Diffusion One model:

```
hf download Mitsua/mitsua-diffusion-one \
  --local-dir ~/Models/mitsua-diffusion-one
```

Now edit the **astro-horse35.py** file to load **mitsua-diffusion-one** instead of **stable-diffusion-v1-5**, then run the code:

```
python astro-horse35.py
```

And you will find an image called something like **output-20260615163224.png**. Compare it with the image created by Stable Diffusion. Although this runs quite quickly, it requires a significant amount of memory. On a Raspberry Pi 5 with 8 GB of RAM, we exhausted all the physical RAM and the swap file, and the operating system stopped responding (we suggest 16 GB for this model). **Figure 4-11** shows a difference in image quality compared to Stable Diffusion models. This is due to Mitsua Diffusion One's more limited training dataset.

Figure 4-11   An image created with Mitsua Diffusion One

You'll find different models create different results based on your prompt, the quality and quantity of the original training data, how many inference steps you take, and so on. A larger model does not necessarily mean better image quality when run in 8 GB or 16 GB RAM. Sometimes it's better to use a model trained to run in a smaller environment. Have fun finding models on Hugging Face Hub and experimenting with them.

Go big or go home with SDXL

It is possible to run Stable Diffusion XL (SDXL) on Raspberry Pi although this has more storage demands (around 72 GB) and takes much longer to produce images. It does, however, produce 1024×1024 resolution images as standard and image quality is much higher.

Fortunately, there's a space-efficient alternative to **hf download**. If you refer to the full path of the model when you call **from_pretrained()**, your code will download only the parts of the model needed. Instead of putting the files in ~/**Models**, it stores them in ~/.**cache/huggingface**. Using this method, the local model files only take up about 14 GB, rather than 72 GB. The first time you run it, you'll need to be connected to the internet, but after that, Python can load the model from the cache.

Let's test out our stable diffusion XL model with the following code (**astro-horse-SDXL.py**):

```python
import os # We need this to expand ~ to your actual home dir
from diffusers import StableDiffusionXLPipeline
import torch # we need this to set the float

# Load the model
model_name = "stabilityai/stable-diffusion-xl-base-1.0"
pipe = StableDiffusionXLPipeline.from_pretrained(model_name,
                            torch_dtype=torch.float32,
                            low_cpu_mem_usage=True)
```

```python
pipe = pipe.to("cpu") # Move the model to the CPU
prompt = "an astronaut riding a horse" # Set the prompt
image = pipe(prompt, num_inference_steps=5,
            width=1024, height=1024).images[0]

image.save("astro-horse-SDXL.png")
```

With just five inference steps, the code will take around 20 minutes to run on a Raspberry Pi 5 with 16 GB RAM where each step took around four minutes. Expect a 35-step image production to take at least a couple of hours. We could have included a negative prompt, but we chose not to as an experiment, to see what the model did with a little more freedom (see **Figure 4-12**).

With 8 GB RAM, it took about 30 minutes. It took even longer without the active cooler (about 40 minutes), because the Raspberry Pi CPU was throttling the whole time. You can monitor whether the CPU is throttling with the command **vcgencmd get_throttled**. If the return value is **0x0**, the CPU is not throttled.

Figure 4-12   An image of an astronaut on a horse generated using the full version of Stable Diffusion XL on Raspberry Pi 5

What's changed?

We had to make a few changes to the code to make it work with Stable Diffusion XL. First, we import and use `StableDiffusionXLPipeline` rather than `StableDiffusionPipeline`. We also had to add `torch_dtype=torch.float32` to our pipe. Stable Diffusion v1.5 uses float32 by default, but more recent models, including XL, use float16. This is smaller and more efficient but is intended to work with GPU (graphics cards) that can support 16-bit model weights. This won't work on the Raspberry Pi CPU, so it must be set specifically to 32-bit, which is slower than running in the model's native format.

Finally, our image is a bit bigger than previous ones: 1024 by 1024 pixels.

Chapter 5

Run a local LLM

Create a GPT Chatbot that runs locally on Raspberry Pi

Large language models (LLMs) are the public face of AI. While there are linear regression models like OLS (ordinary least squares), decision trees like CART (classification and regression), and CNNs (convolutional neural networks) like MobileNet, few people outside the fields of AI and machine learning have heard of such technologies.

It's safe to say that most people working in technology (or technology-adjacent) fields are familiar with the *generative pre-trained transformer (GPT)*, or at the very least, the chat application built on top of a GPT model: ChatGPT. GPT is a large language model architecture built on top of deep learning models called *transformers*.

So welcome to the public face of AI. In this chapter we are going to explore running large language models on Raspberry Pi hardware. We'll start by exploring various LLM models running on a Raspberry Pi 5, first from the command line, and then using the Open WebUI interface (which brings a ChatGPT-like interface to our locally running LLM).

Once this is up and running you can access the web interface from any other computer on your network, dedicating your Raspberry Pi's resources to running the LLM.

Later in the chapter we switch to using the Raspberry Pi AI HAT+ 2 and build a chatbot running on a Raspberry Pi 5 fitted with AI HAT+ 2. This leaves our Raspberry Pi 5 CPU and RAM free to perform other compute tasks, while the LLM runs on the AI HAT+ 2 board.

> **WHAT YOU'LL NEED**
>
> - Raspberry Pi 5 8GB (or 16GB)
> - Internet connection
> - AI HAT+ 2 (optional)
> - 64 GB or more storage
> - AI HAT+ 2 documentation:
>   **rpimag.co/aihatdocs**
>   **rpimag.co/aisoftware**

It is possible to run an LLM on Raspberry Pi hardware, even on a Raspberry Pi 4 with just 4GB of RAM. Albeit a very small and limited model. For more serious workloads you'll need a Raspberry Pi 5 with at least 8 GB of RAM. These enable you to run larger models with up to 7-billion parameters. A 16 GB model will support even larger models, or alternatively, support smaller models while leaving some RAM free for other compute tasks. As will the AI HAT+ 2, which lets you run LLMs and vision processing tasks, and even combine the two with vision-language models (VLMs).

On a Raspberry Pi 5 with 8GB RAM you will be pleasantly surprised by the results. The good news is that setting up an LLM on Raspberry Pi is remarkably easy.

From install to Hello, LLM

We are going to use an open source platform called Ollama (Omni-Layer Learning Language Acquisition Model). Ollama makes it easy to download, install, and run AI language models.

Open a Terminal window (or SSH into your Raspberry Pi) and run the following command:

```
curl -fsSL https://ollama.com/install.sh | sh
```

This will install Ollama to **/usr/local**. Confirm that it is installed with:

```
ollama --version
```

Confirm that the service is running (**Figure** 5-1) with (press Q to exit):

```
systemctl status ollama
```

```
File  Edit  Tabs  Help
lucy@aipi:~ $ curl -fsSL https://ollama.com/install.sh | sh
>>> Installing ollama to /usr/local
>>> Downloading ollama-linux-arm64.tar.zst
######################################################################## 100.0%
>>> Adding ollama user to render group...
>>> Adding ollama user to video group...
>>> Adding current user to ollama group...
>>> Creating ollama systemd service...
>>> Enabling and starting ollama service...
Created symlink '/etc/systemd/system/default.target.wants/ollama.service' → '/et
c/systemd/system/ollama.service'.
>>> The Ollama API is now available at 127.0.0.1:11434.
>>> Install complete. Run "ollama" from the command line.
WARNING: No NVIDIA/AMD GPU detected. Ollama will run in CPU-only mode.
lucy@aipi:~ $ ollama --version
ollama version is 0.30.2
lucy@aipi:~ $
```

Figure 5-1   Check that Ollama is installed and running correctly

TRY TURNING IT OFF AND ON AGAIN

You can stop and start the Ollama service with:

```
sudo systemctl stop ollama
```

```
sudo systemctl start ollama
```

Install your LLM

Let's install our first LLM. You can do this with the `pull` option. Enter:

```
ollama pull llama3
```

This connects to Ollama's model registry and downloads the files from the `llama3:latest` model into your Raspberry Pi's Ollama model cache.

Start your LLM

You can start a chat with the Llama3 model like this:

```
ollama run llama3
```

Type a question and press **ENTER** and it will generate a response:

```
>>> explain quantum computing to me like I'm a five year old
Oh boy, are you ready for something cool?

So, you know how we use computers to play games, watch videos,
and talk to our friends? Well, those computers work with
special things called "bits" that can be either 0 or 1. It's
like a light switch - it's either on (1) or off (0).

But imagine if you had a special kind of toy box that could
hold lots and lots of tiny lights that could be BOTH ON AND
OFF AT THE SAME TIME! That would be really weird, right?

That's basically what quantum computers are. They use special
"qubits" (quantum bits) that can be many things at the same
time, like 0, 1, or even both 0 and 1 at once! It's like a
magic light switch that can change color all by itself.
```

It takes a little while to generate the response (see **Figure 5-2**). When the LLM has finished you can ask another question. There are many caveats with using LLMs locally on your machine. Unlike online services such as ChatGPT and Perplexity, your local model does not have access to the web, so its information will be limited to a cut-off date (the date the model was trained).

You can often ask an LLM to tell you this date:

```
>>> what is your cutoff date
I'm a large language model, my training data was up to 2022.
My knowledge cut off is December 31, 2022. This means that I
have information and training based on events, research, and
developments up until that point in time.
```

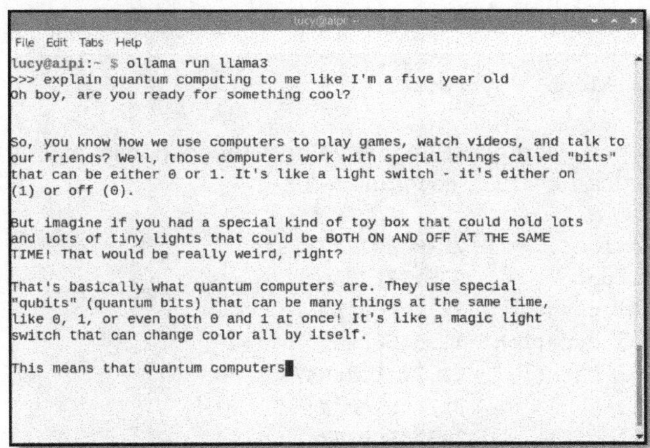

Figure 5-2 Asking the LLM a question

You are also limited to entering text. The LLM will not work with images, files, or other data sources. And like all LLMs it is prone to hallucinations so take what it says with a pinch of salt and be sure to cross-reference important information.

When you are finished with asking questions of your LLM you can return to the command line with: **/bye**.

CHECK THE OPTIONS

Enter **/?** at the prompt to see all the other options available to you.

Verbose mode

Now that you have had a chance to look at your first model, it's time to look at what other LLM models are available to you. There are plenty of models available in Ollama and even more available online using services like Hugging Face.

Before you start to explore models it's a good idea to get some details on the performance of what you are using. For this you'll want to start your model in verbose mode, like this:

```
ollama run llama3 --verbose
```

Now when you ask a question you will get the response, and at the end some statistics, like this (see **Figure 5-3**):

```
total duration:        2m40.007712819s
load duration:         536.722253ms
prompt eval count:     22 token(s)
prompt eval duration:  1.956683s
prompt eval rate:      11.24 tokens/s
eval count:            317 token(s)
eval duration:         2m37.50936s
eval rate:             2.01 tokens/s
```

Here is what these statistics mean:

`total duration`

> The entire time from the moment you ran the command to when Ollama sent the finished output.

`load duration`

> How long it took Ollama to load the model into memory.

`prompt eval count`

> This is how many tokens your prompt contained (in our case 22).

`prompt eval duration`

> How much time Ollama spent reading and processing the 22 tokens.

---

prompt eval rate

The rate at which the tokens were processed. Smaller models process tokens more quickly.

eval count

How many tokens Ollama generated in response.

eval duration

The time it took to generate the tokens.

eval rate

The model's generation speed. This was about 2 tokens per second on our test Raspberry Pi 5 with 8 GB RAM.

Armed with this information you can start to compare models and decide what you want. All models are a trade-off between speed and size. The larger the model, the more accurate (or knowledgeable) it is; but the eval rate with be correspondingly slower.

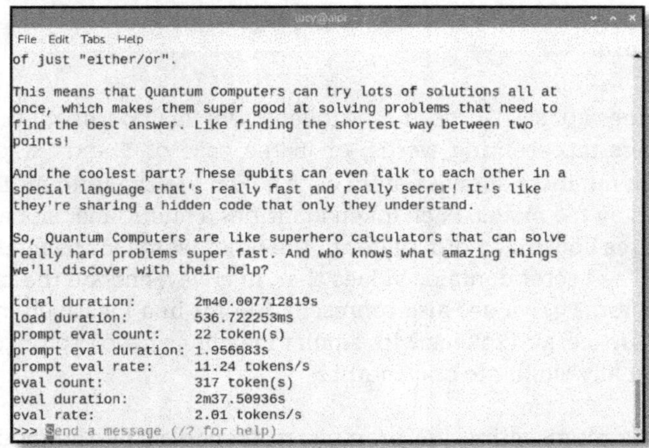

Figure 5-3 Looking at LLM statistics in verbose mode

What is a token?

LLMs break up your prompt into *tokens* using a *tokenizer*. A token usually represents a meaningful chunk of text. This includes spaces and special characters. It also includes metadata tokens, so the total number of tokens reported as the `prompt eval count` shown earlier should exceed the number of words in your prompt.

For example, our prompt "explain quantum computing to me like I'm a five year old" is broken up by the tokenizer and processed like this:

```
95444: 'explain'
31228: ' quantum'
25213: ' computing'
  311: ' to'
  757: ' me'
 1093: ' like'
  358: ' I'
 2846: "'m"
  264: ' a'
 4330: ' five'
 1060: ' year'
 2362: ' old'
```

The LLM does not see words. It sees tokens, which, underneath the hood, are numbers representing words, or in the case of `I` and `'m`, parts of words. The numbers shown before each token are the unique ID of the token used by the model. Each token identifies a multidimensional *vector* that captures how that word relates to the other words in the model's vocabulary. The vector contains values that, in a way, encode the meaning of that vector. The values also represent a point in a massively multidimensional space, and tokens with similar meanings have vectors that are mathematically similar to one another.

The vectors representing the prompt are processed by the model. The model's response is to generate a token based on the most likely next token and continues generating tokens until it predicts an end-of-sentence token or hits a stop condition.

Look at the model

The Llama3 model runs on Raspberry Pi 5, but at two tokens per second, you'd be hard pushed to call it speedy. Fortunately, a lot of work has been done on models since then, and you can get a great speed boost by checking out some of the more recent models.

First let's look at some model information. Back in Terminal, enter:

```
ollama show llama3
```

This will show model information (see also **Figure 5-4**):

```
Model
    architecture          llama
    parameters            8.0B
    context length        8192
    embedding length      4096
    quantization          Q4_0
```

Let's break this down:

architecture (llama)

    This is the family tree. In this case, Meta's LLaMa design.

```
File  Edit  Tabs  Help
lucy@aipi:~ $ ollama show llama3
  Model
    architecture        llama
    parameters          8.0B
    context length      8192
    embedding length    4096
    quantization        Q4_0

  Capabilities
    completion

  Parameters
    num_keep    24
    stop        "&lt;|start_header_id|&gt;"
    stop        "&lt;|end_header_id|&gt;"
    stop        "&lt;|eot_id|&gt;"

  License
    META LLAMA 3 COMMUNITY LICENSE AGREEMENT
    Meta Llama 3 Version Release Date: April 18, 2024
    ...

lucy@aipi:~ $ 
```

Figure 5-4 Model information

parameters (8.0B)

The number of weights inside the model. In this case we have 8 billion. Having more parameters means a richer internal world, but slower speed. Our Raspberry Pi just about runs but is occasionally a bit wheezy. Having fewer parameters means faster speeds.

context length (8192)

This is the model's working memory. The model can keep 8192 tokens (roughly six thousand words) in memory. Anything older than that falls off its mental chalkboard.

embedding length (4096)

This is the depth of meaning assigned to a chunk of text. It converts the token into a 4096-dimensional space. Deeper embeddings can encode more meaning, which can create the impression of more expressive thought (though remember, the model does not "think").

quantization (Q4_0)

> This is where the models' numbers are compressed to save memory. Q4_0 means each weight is stored in 4-bit precision (from its original floating point during training). Quantization makes the model faster at the cost of some precision.

The Llama3 model has a high parameter value, and we could run a much faster, and more modern model to get similar results at a higher speed. So, let's see what newer models are available.

Ollama library

Open your web browser and head to **ollama.com/search**. Here you'll find a library of models with a default list of the most popular **Figure 5-5**.

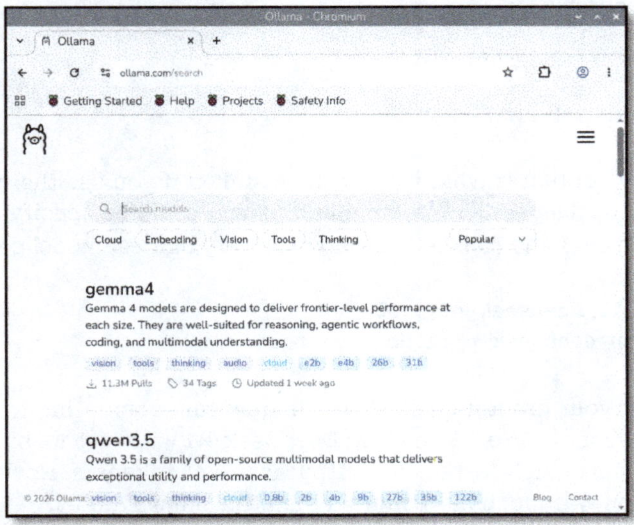

Figure 5-5   Ollama's model library

Pay attention to the number of parameters next to each model. The popular gpt-oss model by OpenAI is only available in 20b and 120b versions (both will be achingly slow).

We should stick with models that are 8b or less. Let's look at one of the lightest modern models around: "DeepSeek-R1". Search for it (or find it in the list) and click to get more information, including a list of available quantizations of the model.

You'll typically see models like:

- deepseek-r1:latest

- deepseek-r1:1.5b

- deepseek-r1:7b

- deepseek-r1:8b (latest)

- deepseek-r1:14b

- deepseek-r1:32b

- deepseek-r1:70b

- deepseek-r1:671b

The "latest" option is what is typically installed if you use the standard `ollama pull deepseek-r1` command. But you can also specify specific versions to `pull` and `run`. Let's see how the very lightest model performs:

```
ollama pull deepseek-r1:1.5b
ollama run deepseek-r1:1.5b --verbose
```

Now enter your prompt again: "explain quantum computing to me like I'm a five year old". You'll find that DeepSeek-R1 with just 1.5 billion parameters runs much faster. Our output was 8.43 tokens/s. However, its explanation was much more limited than the Llama3 model.

After some time spent showing us its deliberation process, it said:

```
In summary, quantum computing offers a fundamentally different
approach to processing information using qubits and
entanglement. While still theoretical and challenging, it has
the potential to solve certain problems exponentially faster
```

than classical computers, though building and maintaining
large-scale systems remains an ongoing goal.

Which doesn't quite hit our brief of "like I'm a five year old". We can
test out other models using **pull** and **run** such as deepseek-r1:8b and
phi3:3.8b. These offer a great combination of speed and functionality.

> **MODELS TO EXPLORE**
>
> Here are some large language models to explore with Ollama:
>
> - DeepSeek-R1 (1.5b, 7b)
> - Phi3 (3.8b)
> - Llama3 (8b)
> - Llama3.2 (1b, 3b)
> - TinyLlama (1.1b)
> - Gemma 2 (2b)
> - Qwen 2.5 (3b)

Manage and remove models

LLMs are large and memory intensive. When testing them out you may
find yourself quickly filling up your storage space. To see your installed
models run:

```
ollama list
```

Remove unwanted models with **ollama rm**. To remove the llama3 model
we installed at the start, enter:

```
ollama rm llama3
```

Use **ollama list** again to confirm that the model has been removed.

Check your storage

Use **df -h** to check how much storage space you have free on your SD card or SSD drive. This command lists usage for all the filesystems mounted by Raspberry Pi OS. Check for the item mounted on the root of your drive, indicated by **/**. Here it is mounted on **/dev/mmcblk0p2**:

```
Filesystem         Size  Used Avail Use% Mounted on
udev               7.9G     0  7.9G   0% /dev
tmpfs              3.2G   16M  3.2G   1% /run
/dev/mmcblk0p2      28G   21G  6.1G  78% /
tmpfs              8.0G  688K  8.0G   1% /dev/shm
```

Looking at the **Avail** and **Use%** columns we can see that we have 6.1GB of storage space available on our drive, and it is 78% full.

Introducing Open WebUI

Interacting with chatbots from the command line soon gets tiring. The good news is that you can connect Ollama to a web interface called Open WebUI (**openwebui.com**). Open WebUI provides us with a Chat GPT-like web interface with interactive chats, bookmarks,

> **OPEN WEBUI DOCUMENTATION**
>
> Bookmark the Open WebUI documentation website: **docs.openwebui.com**

Open WebUI needs to run in a Docker container. This is because Open WebUI is incompatible with Python 3.13 (the version of Python included with Raspberry Pi OS Trixie). Docker provides a *containerised* environment for stable operation. "Set up Docker" on page 24 explains how to install Docker on your Raspberry Pi.

First, make sure that Ollama is running:

```
systemctl status ollama
```

Press **Q** to exit systemctl. Now run Open WebUI in a docker instance:

```
docker run -d -e OLLAMA_BASE_URL=http://127.0.0.1:11434 \
  -v open-webui:/app/backend/data --name open-webui \
  --network=host --restart always \
  ghcr.io/open-webui/open-webui:main
```

Let's break down this Docker command so we can better understand
what's going on:

docker run -d

runs the container 'detached' in the background.

-e OLLAMA_BASE_URL=http://127.0.0.1:11434

sets the environment variable to point to our localhost (127.0.0.1)
and the Ollama API Protocol port running on 11434.

-v open-webui:/app/backend/data

creates a volume for our Docker container called **open-webui**.

--name open-webui

sets the name of our Docker container.

--network=host

makes the container use the host machine's network so it can access
port 11434.

--restart

always restart the container if it crashes.

ghcr.io/open-webui/open-webui:main

gets the container image from GitHub.

Open WebUI uses port 11434 to communicate with the Ollama instance
that's running on your Raspberry Pi (the host machine). This communi-
cation follows a *protocol* (in this case, the Ollama API Protocol), which
defines the ground rules for communicating with Ollama. Open WebUI,
running in the Docker container, offers a web application that runs on
your computer and listens for requests on port 8080.

After it's done pulling the Docker files and returns to the command prompt, wait a minute or two to give Open WebUI time to start up. Next, open a web browser and connect to the Open WebUI interface at this URL: `http://localhost:8080`

You will see a welcome page for Open WebUI **Figure 5-6**. Click the **Get Started** arrow. The first user to access Open WebUI will need to create an admin account (**Figure 5-7**). Fill out the **Name, Email,** and **Password** fields and click **Create Admin Account**. You will be logged in and presented with some release notes.

Click **Okay, Let's Go!** to reach the main interface.

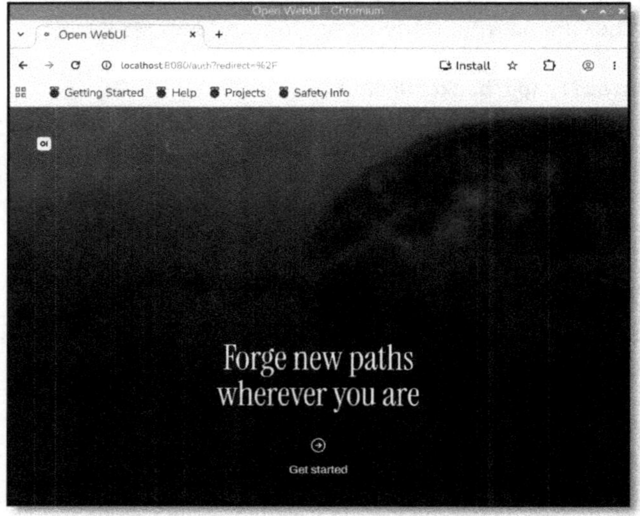

Figure 5-6   The Open WebUI welcome interface

ALTERNATIVE SIGN IN

You can also sign in using the URL specified when we started the Docker instance: http://127.0.0.1:8080 instead of using `localhost`.

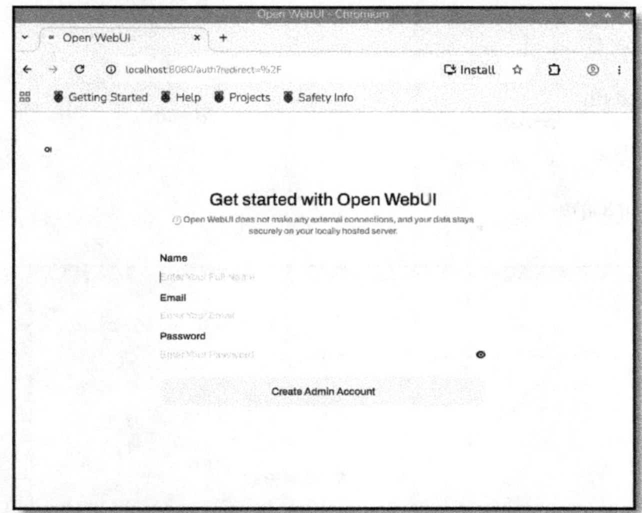

Figure 5-7   Open WebUI admin creation

Using Open WebUI

Before chatting to Open WebUI, you will need to select an LLM to interact with. Click **Select a Model** in the top left of the screen. If you have previously installed models with Ollama, pick a model from the list.

If you didn't install a model earlier ("Install your LLM" on page 84), you will need to pull a model from the Open WebUI interface. "Pull a model" on page 99 explains how to do this.

Now start your chat in the text box marked with "How can I help you today?" as shown in **Figure 5-8**. You can ask questions, have back-and-forth on answers, and it will suggest follow-up questions. It resembles services like ChatGPT (albeit using smaller models with more limited training).

Click the **Open Sidebar** icon at the top left for more options:

- ▶ **New Chat**

- ▶ **Search**

- ▶ **Notes**

- ▶ **Workspace**

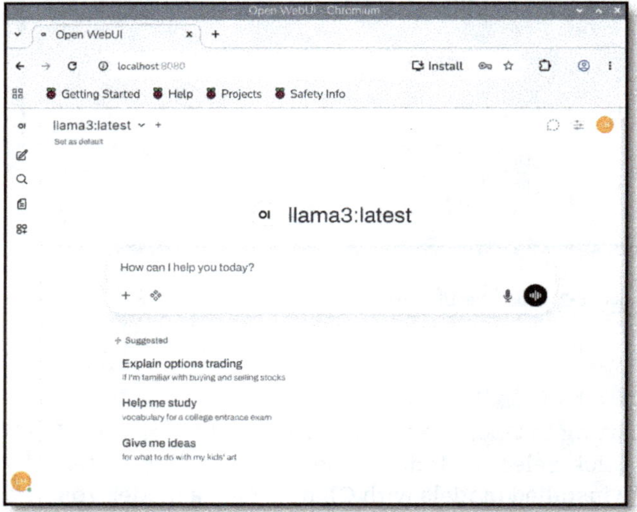

Figure 5-8   The Open WebUI chat interface

You will also see folders and a chat history. Clicking on your account icon in the bottom left reveals further options, including:

- ▶ **Settings**

- ▶ **Admin Panel**

- ▶ **Archived Chats**

- ▶ **Playground**

- ▶ **Sign Out**

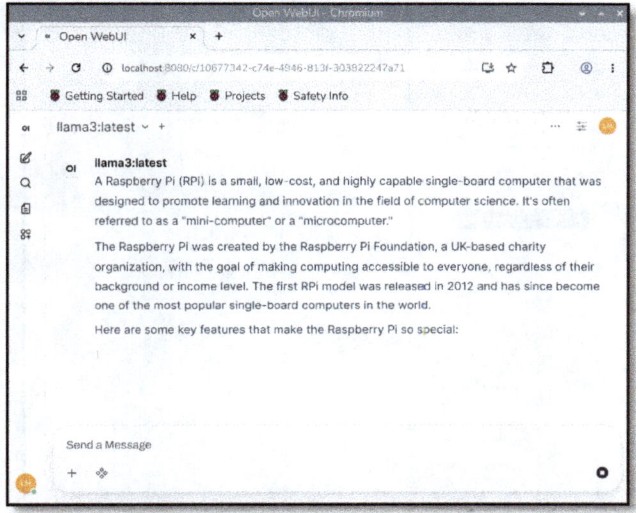

Figure 5-9   Asking your LLM questions

Figure 5-9 shows an active chat session. Clicking on the account icon in the top right provides the same settings plus **Documentation, Releases, and Keyboard shortcuts.** There is also a **Controls** icon to the left of your account icon. This reveals a list of advanced parameters for you to tweak. For example, adjusting the **Temperature** setting makes the model answer more creatively (whereas setting it too low can make it repetitive). Hovering over each setting will provide a help bubble that explains what each setting does.

Pull a model

You can add new models to Ollama from inside the Open WebUI interface. Click the + icon next to the model list dropdown menu. Then click **Select a Model** and begin typing the name of a model (e.g., mistral). Next, click **Pull "mistral" from Ollama.com** and it will download it to the Ollama space (see **Figure 5-10**). You can now select it from the list of models.

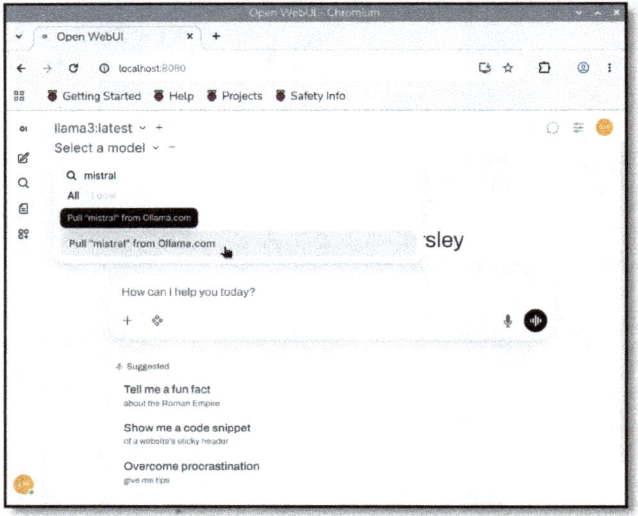

Figure 5-10 Pulling a model from the Ollama library inside the Open WebUI interface

Models can be managed inside Open WebUI by clicking **Settings and Models.** You can shut down the Docker container with `docker stop open-webui`. You can start it again with `docker start open-webui`.

SHARE SPACE

Because your Open WebUI docker instance is connected to Ollama via port 11434, it will add new models to your Ollama package running on Raspberry Pi. You can also add, and remove, LLMs from Ollama on your Raspberry Pi command line and the two will sync in both areas.

Run an LLM on AI HAT+ 2

Now that we have investigated running an LLM on our Raspberry Pi 5, we can look at placing generative models onto our AI HAT+ 2 hardware. This enables us to offload the LLM away from Raspberry Pi, freeing up our computer for other compute tasks.

AI HAT+ 2 features a Hailo-10H neural processing unit (NPU) capable of delivering 40 TOPS of performance, alongside 8 GB of dedicated RAM for running LLMs.

Like Ollama you can access it via the command line or using an Open WebUI interface. And you can expect faster performance. However, the downside is that only models optimised by Hailo work with the Hailo-10H NPU, so you are limited in model choice. They have selected a great choice of models though, capable of function calling and code generation.

AI HAT+ 2 software prerequisites

Before running vision AI models or GenAI models on your Raspberry Pi 5, you must configure the required software. Broadly, this consists of the following tasks, performed in order:

1. Update Raspberry Pi OS. This ensures that your Raspberry Pi OS packages are fully up to date.

2. Install the necessary software dependencies that allow the operating system (OS) and applications to detect, communicate with, and run AI models on the Hailo NPU.

3. Reboot and verify that your AI hardware is correctly detected and ready to use.

Ensure that the Raspberry Pi 5 is running Raspberry Pi OS Trixie with the latest software installed, and that it has the latest Raspberry Pi firmware:

```
sudo apt update
sudo apt full-upgrade -y
sudo rpi-eeprom-update -a
sudo reboot
```

After updating your Raspberry Pi with the latest Raspberry Pi software and firmware, the following dependencies are required to use the NPU:

- ▸ The Hailo kernel device driver and firmware.

- ▸ Hailo RT middleware software.

- ▸ Hailo Tappas core post-processing libraries.

To install the required dependencies for AI HAT+ 2, open the Raspberry Pi Terminal and run the following commands:

```
sudo apt install dkms
sudo apt install hailo-h10-all
```

After installing the dependencies, you must reboot your Raspberry Pi 5. You can do this from the Raspberry Pi Terminal using this command:

```
sudo reboot
```

When your Raspberry Pi 5 has finished booting back up again, run the following command to check that everything is running correctly:

```
hailortcli fw-control identify
```

You can find complete documentation at **rpimag.co/aihatsoftware**. Be sure to look for the AI HAT+ 2 instructions there rather than the instructions for the AI Kit or the original AI HAT+.

Start the Hailo Ollama server and run LLMs

The next step is to download the Hailo Model Zoo Debian package:

```
wget https://rpimag.co/hailo_gen_ai_model_zoo.deb
```

That URL will download a version of the Model Zoo package that works with the examples in this chapter. As of this writing, that's **hailo_gen_ai_model_zoo_5.1.1_arm64.deb**, but it will be saved on your computer with the filename shown in the URL.

Next, Install the Debian package:

```
sudo dpkg -i hailo_gen_ai_model_zoo.deb
```

After everything is installed, you'll need to start the local `hailo-ollama` server to expose a REST API for LLM requests, and then download and run some LLMs. In a Raspberry Pi Terminal, run the following command to start the local hailo-ollama server:

```
hailo-ollama
```

Open a new Terminal window, then run the following command to get a list of LLMs:

```
curl --silent http://localhost:8000/hailo/v1/list
```

> **PORT 8000**
>
> Note that hailo-ollama uses a different port than the Ollama server running on our Raspberry Pi CPU, which you installed earlier in this chapter. It uses 8000 instead of 11434. This ensures that the two versions of Ollama do not interfere with each other.

Run the following command to download a model from the provided list, replacing `examplemodel:tag` with any listed model:

```
curl --silent http://localhost:8000/api/pull \
    -H 'Content-Type: application/json' \
    -d '{ "model": "examplemodel:tag", "stream" : true }'
```

For example, you could download `qwen2:1.5b` like so:

```
curl http://localhost:8000/api/pull \
    -H 'Content-Type: application/json' \
    -d '{ "model": "qwen2:1.5b", "stream" : true }'
```

Run the following command to send a query to the LLM with a POST request, replacing **examplemodel:tag** with whichever model you've downloaded and want to run, and **prompt** with your prompt.

```
curl http://localhost:8000/api/chat \
    -H 'Content-Type: application/json' \
    -d @- <<EOF
{"model": "examplemodel:tag",
 "messages": [
   {"role": "user",
    "content": "prompt"
   }]}
EOF
```

The **-d @-** argument tells curl to get its POST request from *standard input*. Standard input is the name of the input stream used for text you type into a running program.

The lines that start with **<<EOF** and end with **EOF** are a *here-document*, a kind of shell syntax that lets you redirect multiple lines of text into standard input. The lines between **<<EOF** and **EOF** make up the entire POST message and are a bit easier to read than if we tried to cram all that between single quotes.

Here's how you would ask qwen2:1.5b to translate a sentence to French:

```
curl http://localhost:8000/api/chat \
    -H 'Content-Type: application/json' \
    -d @- <<EOF
{"model": "qwen2:1.5b",
 "messages": [
   {"role": "user",
    "content": "Translate to French: The cat is on the table."
   }]}
EOF
```

If you don't feel like typing all that in, you can find a shell script with this code in this book's GitHub repository (see *Welcome* on page v). you can use **git clone** to check it out, then use the **cd** command to change to the **ai-projects-raspberrypi/ch05** subdirectory, as in:

```
cd ~/ai-projects-raspberrypi/ch05
```

Then run this command with `./qwen_translate.sh`. The output should look similar to this:

```
{"model":"qwen2:1.5b","created_at":"2026-02-03T13:34:02.309Z",
 "message":{"role":"assistant","content":"Le"},"done":false}
{"model":"qwen2:1.5b","created_at":"2026-02-03T13:34:02.434Z",
 "message":{"role":"assistant","content":" chat"},"done":false}
{"model":"qwen2:1.5b","created_at":"2026-02-03T13:34:02.557Z",
 "message":{"role":"assistant","content":" est"},"done":false}
{"model":"qwen2:1.5b","created_at":"2026-02-03T13:34:02.680Z",
"message":{"role":"assistant","content":" sur"},"done":false}
{"model":"qwen2:1.5b","created_at":"2026-02-03T13:34:02.803Z",
 "message":{"role":"assistant","content":" la"},"done":false}
{"model":"qwen2:1.5b","created_at":"2026-02-03T13:34:02.926Z",
 "message":{"role":"assistant","content":" table"},"done":false}
{"model":"qwen2:1.5b","created_at":"2026-02-03T13:34:03.0483Z",
 "message":{"role":"assistant","content":"."},"done":false}
{"model":"qwen2:1.5b","created_at":"2026-02-03T13:34:03.171Z",
"message":{"role":"assistant","content":""},"done":true,
"done_reason":"stop","total_duration":1178151410,"eval_count":7}
```

Look closely at the output and you will see:

```
"content":"Le",
"content":" chat",
"content":" est"}
"content":" sur"}
"content":" la"}
"content":" table"}
```

Along with other information, such as the timestamp, model, and whether each token has hit a stop point (`"done":false` and `"done":true,"done_reason":"stop"`). The final output also contains the duration and the token count.

While this information is interesting for development purposes, it is less fun as a practical user. For that we will connect our Hailo-Ollama instance to Open WebUI. If you created an Open WebUI Docker container earlier, you may want save disk space and confusion by removing it. (You can keep if it you'd like, though). First, check for the Docker instance. In a terminal window, run:

```
docker ps -a
```

If you see a Docker instance named **open-webui**, remove it with:

```
docker stop open-webui
docker rm open-webui
```

Next, check and remove its connected volume:

```
docker volume ls
docker volume rm open-webui
```

Confirm that both are removed:

```
docker ps -a
docker volume ls
```

Install Open WebUI

If you didn't install Open WebUI earlier (see "Introducing Open WebUI" on page 94), you will need download the Open WebUI image required to run the frontend layer. If you already have the Open WebUI image it will update it instead. So, it doesn't hurt to run this step again:

```
docker pull ghcr.io/open-webui/open-webui:main
```

Ensure that **hailo-ollama** is already running. If you don't have it running in another Terminal session, open a new Terminal and run:

```
hailo-ollama
```

Open another terminal window and start the Open WebUI container and connect it to the hailo-ollama backend server:

```
docker run -d -e OLLAMA_BASE_URL=http://127.0.0.1:8000 \
  -v open-webui2:/app/backend/data --name open-webui2 \
  -e PORT=8888 --network=host --restart always \
  ghcr.io/open-webui/open-webui:main
```

> **SPOT THE DIFFERENCES**
>
> Note that this command differs slightly from our previous Ollama docker container. First, it is connected to the Hailo Ollama instance on port **8000** instead of regular Ollama on port **11434**. Second, we're using a different name for the container and volume (**open-webui2** rather than **open-webui**). Finally, we configure the Open WebUI web server itself to listen on port 8888 instead of the default (8080).

The container can take up to a minute to initialise. To view progress and logs, run the following command and then wait until the logs indicate that the server is running and ready to accept connections.

```
docker logs open-webui2 -f
```

Access Open WebUI in a web browser and enter the following URL: **http://127.0.0.1:8888**. This opens a chat interface where you can select a model and begin interacting with the LLM.

As in "Introducing Open WebUI" on page 94, you will need to create an admin account on your first login. The difference here is that you can only pull and use the models supplied by Hailo.

If you pulled models in the command line they will appear in the web interface, otherwise you will need to enter their name in the **Search a model** field. At the time of writing, the models available are:

Model	Purpose
deepseek_r1_distill_qwen:1.5b	Logic and mathematics
llama3.2:3b	All round chat and text
qwen2.5-coder:1.5b	Code generation and programming tasks
qwen2.5-instruct:1.5b	User commands and instructions
qwen2:1.5b	Backwards compatibility with older Qwen

For our chatbot we suggest you try Llama 3.2. Click the **+** that appears next to the current model name, click **Select a model**, and enter llama3.2:3b into the search field. Next, click the **Pull** option to download it to our `hailo-ollama` installation. When it's installed, click the current model and choose Llama 3.2 instead.

Network access

It is possible to access the LLM on your Raspberry Pi across a local area network (LAN) via Ethernet or Wi-Fi. With a Raspberry Pi positioned on your local network you can set it up as a local LLM that you can access via the Open WebUI interface.

You will need to know your hostname. Open Terminal and enter:

```
hostname
```

Ours is `aipi`. Now switch to a web browser on another computer on your same network. Enter the hostname followed by `:8888` to access our Open WebUI port on the network:

```
http://<hostname>:8888
```

On our machine this is `http://aipi:8888`

> ### TRY .local AND 127.0.0.1
>
> On some networks you may need to append `.local` to your hostname, like this:
>
> ```
> http://aipi.local:8888
> ```
>
> If you can't connect using a hostname, run the command `ip address show` in a Terminal window on your Raspberry Pi and look for an IP address under either **eth0** (if you are connected via Ethernet) or **wlan0** (Wi-Fi). The IP address will appear immediately after the text **inet**. Try using that IP address instead of the hostname, as in `http://192.168.1.242:8888`.

You will be presented with the login screen. Enter the name and password you created for the admin user. Now you will be able to interact with your chatbot on the local network just like using a remote web GPT service. All your conversations remain on the local network.

Open WebUI is also optimised for web browsers on mobile phones. So, you can access using a web browser on your phone.

> ### MAKE AN APPLICATION
>
> On iOS and Android phones you can turn the web page into an application. On Android you can choose **Settings, More,** and **Add to Home Screen**. In iOS click **Share** and **Add to Home Screen**.

Next steps

We hope you have enjoyed creating a chatbot for your Raspberry Pi. The models are much smaller than some of the internet-based options. However, they are perfectly able to answer questions and work on text. We find the coder models especially helpful for debugging our Python code. And, crucially, you are no longer exposed to sharing critical information with third-party companies and having them train their models on the files and text you provide them.

We'll revisit chatbots in Chapter 7, *Speech to text and back again*, where we'll build a voice-powered chatbot. That chapter takes a different approach, and will show you how to write chatbots from the ground up using Python.

Chapter 6

Training models on sensor data

In which a microcontroller teams up with a computer for real-world machine learning

There is a large market for hugely powerful machine learning accelerators that boost the processing power of computers to allow them to train and run large models. However, there is another end to the market. *TinyML* enables the same ML processes, stripped down to run on microcontrollers like the Raspberry Pi RP2350 chips found in Raspberry Pi Pico 2 micro-controller boards.

This chapter's project uses both a Raspberry Pi and a Pico 2. For the microcontroller, it must be a Pico 2 or other RP2350 board as the machine learning library we'll use isn't (at the time of this writing) compatible with the RP2040 chip found in the original Raspberry Pi Pico.

We're going to make a magic wand that can detect gestures and light up when it's flicked. We'll do this using an accelerometer that can detect movement and stream the data through a machine learning model that we've trained. When it detects a flick, we'll sparkle the LEDs.

The first thing we need is a Pico 2 set up with MicroPython and Blinka. This Python library enables us to use CircuitPython libraries and thereby get easy access to a lot of sensors. The specific sensor we'll use is the *LSM303AGR*, a 3-axis accelerometer and 3-axis magnetometer. Since you'll be training the model yourself, you should be able to swap this out for just about any other accelerometer that you can read in MicroPython, and it should work.

We need some data to train our model, and to get this, we'll hook up our sensor to our Pico 2 and stream readings over USB to our Raspberry Pi where we'll train our model. We also need a way of *labelling* our data; that means highlighting which parts of our data show flicks and which show other bits of movement. To do this, we'll add a button that we can press when we're flicking the wand. You can hook this up on a breadboard or use a different prototyping method.

Bear in mind that the way you arrange your circuit for training must be the same as the arrangement for its final use. Slight variations in weight or shape of the wand will affect the accelerometer readings, so you may want to read through this whole section before running the training (that said, once you've got everything set up, it's pretty straightforward to re-run the training).

WHAT YOU'LL NEED

- Raspberry Pi
- Raspberry Pi Pico 2 or Pico 2 W
- Solderless breadboard
- Push-button switch
- LSM303AGR accelerometer
- 4 socket-to-pin jumper wires
- 2 pin-to-pin jumper wires
- 1 or 2 strips of 10 × WS2812B LEDs
- Soldering iron, solder, and wires

Figure 6-1 shows the wiring diagram for training the magic wand. When training, make sure that you have the accelerometer at the same orientation that you'll have it in your final build.

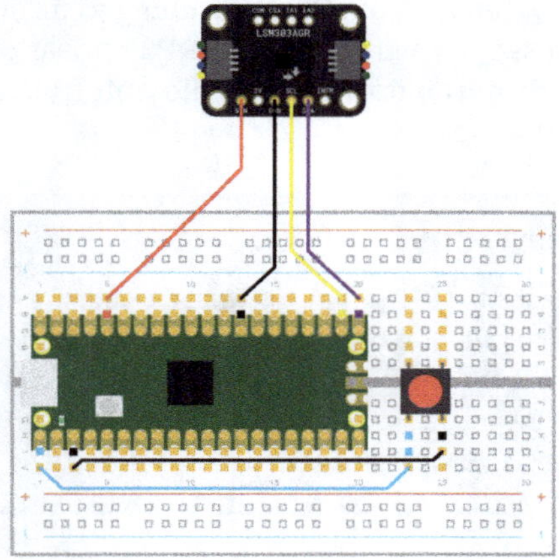

Figure 6-1   The magic wand training circuit

The first part of our program for this project is for our Pico 2 to stream the data over USB serial.

Installing Python and MicroPython

Because you're using a Raspberry Pi with Raspberry Pi OS, you'll already have Python installed.

Microcontrollers, such as Raspberry Pi Pico 2, are a bit different to computers with operating systems because they run one bit of software at a time. Rather than install different programs and choose what to run, you install a bit of firmware that replaces whatever was on the device previously. MicroPython is a full implementation of the Python 3 programming language that runs on embedded hardware like Pico. Download it from **rpimag.co/micropython.**

Make sure that you select the right software for your board, because it is closely tied to the hardware. You will need either Pico 2 or Pico 2 W.

Click the link for your board, and you should download a file named something like **<pico version>-latest.uf2**. To flash this to your board, make sure that your Pico is unplugged from your computer, then hold down the BOOTSEL button on Pico and keep it held down while you plug Pico into your computer via USB. Once it's plugged in, you can release the button. You should see a drive called **RPI-RP2** appear on your computer. Copy or drag-and-drop the downloaded file onto this drive. After a few seconds, it should disappear.

> **? RESET FLASH MEMORY**
>
> Installing MicroPython in the usual way will write the data to the storage on Pico, but occasionally things will go a bit awry with some data on there. If things aren't working as you expect after flashing MicroPython, then you can download **flash_nuke.uf2** from **rpimag.co/picoresetflash**. Flash this to your Pico in the usual way, and then reinstall MicroPython.

It might look like everything has disappeared, but it hasn't. While the drive has gone, there's now a serial connection between Pico and your computer. To access this, open Thonny, click the menu in the lower right, and choose **Configure interpreter**. Select **MicroPython (Raspberry Pi Pico)** and **<Try to detect port automatically>** from the **Port or WebREPL** menu, then press **OK**.

Pico: Install Blinka software

Now let's set up the software. In this chapter, we're using MicroPython (on Pico) and Python (on Raspberry Pi). While MicroPython has support for a lot of devices, the best set of modules for using hardware with Python is the CircuitPython Library Bundle. Even though we're not using CircuitPython, we can still use its libraries. The underlying Python language is the same; they just have a slightly different way of interacting with the hardware. To bridge the gap between CircuitPython libraries and MicroPython or standard Python, we use a *compatibility layer* called Blinka. Blinka allows CircuitPython modules to run on both the Raspberry Pi and the Pico, making hardware integration easier.

We'll start by installing two Adafruit modules, Blinka and PlatformDetect. You can download these from the following links:

- ▸ **rpimag.co/blinkagit**

- ▸ **rpimag.co/platformdetectgit**

Download the Blinka source zip file from the release page, then unzip it somewhere on your computer. This is the full set of code for every board supported by the module. Most of this isn't needed, so we'll get rid of the bits for other platforms. The important bits are in the folder named **Adafruit_Blinka_<version>/src**. Inside this, there are two folders that we're going to clean up:

- ▸ **adafruit_blinka/microcontroller** (not to be confused with the **micro-controller** folder in the top level of the **src** directory itself).

- ▸ **adafruit_blinka/board**

In these two folders, you can delete everything except:

- ▸ **adafruit_blinka/microcontroller/generic_micropython** folder and the **adafruit_blinka/microcontroller/rp2040** folder

- ▸ **adafruit_blinka/board/raspberrypi/pico.py**

Once you've removed those, you can upload the rest to Pico. Open Thonny and make sure you're in the Regular UI mode. If you see the text "Switch to regular mode" in the upper right of the Thonny window, click it, then quit and restart Thonny to enable the new mode.

Click the **View** menu in Thonny and make sure that **Files** is checked. This will open a section with two file panes: one showing the files on your computer and one showing those on Pico.

On Pico, we want to make a **/lib** folder that will store our modules. This isn't strictly necessary but does help keep things a bit tidier in cases like this where we have a lot of modules. Right-click in the Pico files pane, then select **New directory** and name it **lib** (it must be named **lib** exactly or MicroPython won't search it for modules). Once you've created **lib**, double-click it to change into this directory.

Now, on the computer files pane, navigate to where you unzipped Blinka and then into the **src** subdirectory. Select everything in there (click on the top item, then hold down **SHIFT** and click on the bottom item), then right-click to open the context menu and select **Upload to /lib**.

That's Blinka set up; we now need PlatformDetect. Fortunately, this is a bit simpler. Just unzip the directory; then, in the Thonny pane, navigate to the **Adafruit_Python_PlatformDetect-<version>** folder and inside that, select **adafruit_platformdetect**, then right-click and select **Upload to /lib**.

In the REPL (**Shell** pane), you can check everything's worked by running the following command at the >>> prompt:

```
import board
```

If that works, then you've installed Blinka correctly. If you get any errors, you can go back and try and fix them.

BLINKA GUIDE

You can find a complete guide for installing Blinka on MicroPython at **rpimag.co/cplibsonmppico**.

There's one more setup task. Although we've got Blinka up and running, we still need our accelerometer library. This is part of the CircuitPython Library Bundle — a single download where you'll find modules for almost all common hardware. Go to **circuitpython.org/libraries** and download the latest **Python Source Bundle** (the bundles above it, labelled as "Bundle for version Y.x", won't work as they are compiled for CircuitPython, not MicroPython). Unzip this and use Thonny to copy the following to the **/lib** folder on Pico:

- **adafruit_bus_device** (folder)

- **adafruit_register** (folder)

- **adafruit_lsm303_accel.py** (file)

Next, save the following code on your Raspberry Pi Pico 2 in its top-level
directory. You can find this code in the GitHub repository (*Example code*
on page ix) in the **ch06** subdirectory as **stream.py**:

```python
from time import sleep, ticks_ms
import board
import busio
import adafruit_lsm303_accel
import machine

button = machine.Pin(0, machine.Pin.IN, machine.Pin.PULL_UP)

i2c = busio.I2C(board.GP17, board.GP16)
accel = adafruit_lsm303_accel.LSM303_Accel(i2c)

while True:
    reading = accel.acceleration
    print(
        ticks_ms(),
        reading[0],
        reading[1],
        reading[2],
        button.value(),
        sep=",")
    sleep(0.1)
```

The accelerometer produces three bits of data that are the acceleration in the X, Y, and Z dimensions. If the device is stationary, then the reading on this will be the acceleration due to gravity — in other words, it will show which way the device is being held up. However, if you move it around, the acceleration on the device will change with the forces on the device.

Save this to your Pico with the file name **stream.py** and try running it from Thonny. You should see a stream of values appearing:

```
1281606,-0.6501809,-0.1529837,9.714468,1
1281711,-0.6501809,-0.1529837,9.676222,1
1281815,-0.6119349,-0.1529837,9.714468,1
```

If you see the error "ValueError: No I2C device at address: 0x19", you'll need to determine the I2C address of your accelerometer. You can scan an I2C bus to find out what (if any) devices are connected to it. This can be a very useful step in ensuring that your circuit is wired up correctly and everything is working.

We'll use standard MicroPython classes for scanning. In this example (i2c_scan.py) we're using the built-in `machine.I2C` class rather than Adafruit's `busio.I2C` class:

```python
import machine
i2c = machine.I2C(0, scl=machine.Pin(17), sda=machine.Pin(16))

print('Scanning i2c bus...')
```

```python
devices = i2c.scan()

if len(devices) == 0:
    print("No i2c devices found!")
else:
    print(f"Found {len(devices)} i2c devices.")

    for dev in devices:
        print(f"Decimal addr: {dev} | Hex addr: {hex(dev)}")
```

After you have determined the correct address for your accelerometer, edit the **lib/adafruit_lsm303_accel.py** file on your Pico, find the line that reads **_ADDRESS_ACCEL = const(0x19)**, and change the **0x19** to the hex address of your sensor.

Capture input with stream

Quit Thonny before proceeding; this will make sure it isn't using the Pico 2's USB serial port. You'll need to keep your Pico 2 connected to your Raspberry Pi over USB, though.

We now need to run **stream.py** in such a way that we can capture its output to a file. To prepare for this, create a new virtual environment on your Raspberry Pi. Activate the virtual environment and install **mpremote**. You can then use **mpremote** to run the Python program on the Pico 2 and save the data to a file. You can do all this with:

```
python3 -m venv ~/wand
source ~/wand/bin/activate
pip install mpremote
mpremote exec "import stream" > ~/Documents/wand_data.csv
```

The final line runs the script (on your Raspberry Pi Pico) and redirects the output to the local file **wand_data.csv** in the **Documents** folder underneath your home directory. This might seem a bit strange because we're not specifying the full name of the **stream.py** program. The **exec** command executes a Python *command*, not a *script*. The Python command is **import stream**. If **mpremote** can't find your Pico, make sure you aren't connected to it in another app (such as Thonny).

The **import** command is generally used to load modules in your code, but it just runs whatever is in the referenced Python file. These files typically create classes, functions, or anything else we want to use, but we can also use it to run scripts like this. We don't need the **.py** on the end, as Python will add this automatically.

You may have seen Python modules with runnable code preceded by **if __name__ == "__main__"**. That **if** statement prevents the code within it from running, unless you ran the module from the command line or within Thonny. If you just **import** such a module, the code within the **if** statement won't be run. Because **stream.py** doesn't have such an **if** statement, **import stream** runs all the code in the file.

The final part of the command, **> wand_data.csv**, uses what's known as a *redirect*. The **>** tells the shell to send any output of the previous command to a file. In this case, **wand_data.csv**. You can use this to capture any output from any program and save it to a text file.

> **CSV FILES**
>
> Comma-separated values — or CSV — files are a common way to store spreadsheet-like data. They're supported by a wide range of software, so if you want to examine the data, you can open and edit it with a spreadsheet application.
>
> A word of warning: CSV files can get messy if you need to include commas in the data itself. There are workarounds such as quoting and escaping, but there are multiple competing methods for doing this and not all software supports all options. For simple numerical data, though, they're hard to beat for ease and portability.

Nothing will display on the screen, but it should keep running. Wave your wand around and occasionally flick it, being sure to press the button when you do. Start pressing the button just before you flick it and keep it held down until the flick is complete. Rotate the wand and try flicking from different orientations. Hold it still for a little bit, but do so in a few different orientations. A few minutes of waving, resting, and flicking should be enough data with which to train your model. Press **CTRL+C** when you are done.

Now let's look at training the model. First, we need to install the necessary parts on your Raspberry Pi. In the same *venv* (virtual environment) we created before, run:

```
pip install scikit-learn numpy emlearn setuptools
```

We introduced scikit-learn, a popular machine learning library, back in Chapter 3, *An introduction to predictive models*. We'll be using emlearn to run our scikit-learn models on our microcontroller. There's more information on using emlearn in its GitHub repo: **rpimag.co/emlearngh**

The following code will load the data we streamed in from our sensor and use it to train a `RandomForestClassifier`. You can find it in the **ch06** folder from the GitHub repo as **build_model.py**:

```python
import csv
from sklearn.ensemble import RandomForestClassifier
from sklearn.model_selection import train_test_split
import emlearn

current_window = []
X_set = []
y_set = []
window_size = 10
rolling_flick = []
data = []
multiplier=1000


def to_int(data_list):
    output = []
    for datum in data_list:
        output.append(int(float(datum) * multiplier))
    return output

def rolling_av(data, index):
    total = 0
    for item in range(index - window_size, index):
        total = total + int(data[item][4])
```

```python
    if total / window_size < 0.5:
        return 1
    return 0

with open('wand_data.csv') as csvfile:
    flickreader = csv.reader(csvfile, delimiter=',')
    counter = 0
    for row in flickreader:
        data.append(row)

for index in range(10, len(data)):
    predictor = []
    for y in range(0, 3):
        for x in range(-window_size, 0):
            predictor.append(data[index + x][y + 1])
    X_set.append(to_int(predictor))
    y_set.append(rolling_av(data, index))

X_train, X_test, y_train, y_test = train_test_split(X_set,
                y_set, test_size=0.2, random_state=42)
estimator = RandomForestClassifier(n_estimators=20,
                max_depth=15, max_features=7, random_state=42)
estimator.fit(X_train, y_train)

score = estimator.score(X_test, y_test)
print("score is: ", score)

# Convert model using emlearn
cmodel = emlearn.convert(estimator, method='inline')

# Save as loadable .csv file
path = 'flick_model.csv'
cmodel.save(file=path, name='flick', format='csv')
print('Wrote model to', path)
```

Before you run it, copy the **wand_data.csv** file to the same directory as
this script. You can run it with **python build_model.py**. If you get the
error message **pos_label=1 is not a valid label: It should be one
of [0]**, it means that, after applying the rolling average (discussed in a

moment), the data doesn't have any valid flicks in it. Try creating the CSV file again but hold down the button a bit longer for each flick.

Let's look a bit closer at what the preceding program does.

Our input data is a stream of readings, each one taken approximately one tenth of a second apart. The input to our classifier should be a chunk of data and a result of that data. Essentially, we want to create two lists: one is a 2D list that contains the data, and one is a 1D list that contains the result.

The chunk of data we want to send to the classifier is a window of one second's worth of data. Each window starts 0.1 second after the previous one. This is called a *sliding window* and while it can seem a bit wasteful as each reading is processed ten times, it ensures that we don't miss a flick because it takes place half in one window and half in the next.

We need to convert our data into a format that the model wants. Our random forest wants integers, and we have floats. We could just do an integer conversion, but that would lose almost all the data, so instead, we first multiply the value by 1000. This means that we keep the data down to the nearest 0.001, which should be enough.

If your sensor returns integer values, you must change `multiplier` to 1 in both **build_model.py** and **ch06-wand.py**.

With machine learning, you usually have to do some preprocessing on the data, and deciding on the right preprocessing can have a big effect on the final result. The preprocessing here is to group it into a sliding window and convert it to integers. It can be much more complex than this.

Each line in the data array contains 30 entries: the last ten readings for each of the X, Y, and Z axes.

We also need to decide if the particular window includes a flick. We do this by looking at the final item in each line of the CSV file, which is the state of the button. Our window is unlikely to be made up entirely of a flick because most flicks are less than a second. We settled for declaring a window a flick if 50% or more of the window includes a button press.

Note that because we recorded the pure button state in the stream, a 1 is no button press while a 0 is a press. However, in our classifier, we want a 1 to be a 'True' value (i.e. a flick). This is why the test is for less than 0.5 rather than more than.

Now we've amassed all the data, we need to use it to train our classifier. We only have limited memory available on our Pico 2, so we need to keep this classifier as small as possible. We've set the values for **n_estimators** (which is the number of trees), **max_depth** (the size of each tree), and **max_features** (the number of features that are used at each stage of each tree). In normal (that is, non-microcontroller) use, you would often want to leave these out and let scikit-learn pick the right size for your classifier.

There isn't a hard-and-fast way for working out what the values should be. We adjusted them until we found a set of small-ish values that gave us accurate results.

Before we train the model, we use **train_test_split()** to set aside 20% of the data for testing and scoring our model. The variable **X_set** contains all the accelerometer readings, and **y_set** contains the button state. During training, we use **X_train** and **y_train**. For testing, we call the model's **score()** function to run predictions using **X_test**. It then scores the model based on how well the model's predictions align with the true values in **y_test**.

There are different types of machine learning models available in emlearn. They have different characteristics and perform well in different circumstances. The main ML features of emlearn are:

- ▸ Classification with RandomForest/DecisionTree models

- ▶ Classification and on-device learning with the K-Nearest Neighbors (KNN) algorithm

- ▶ Classification with Convolutional Neural Network (CNN), using TinyMaix library

- ▶ Clustering using K-means

If you're interested in taking this further, the documentation is a great place to start: **scikit-learn.org**.

Add LED sparkle

We've now got our model and our hardware; we just need to load it onto our Pico 2 and run it. The only extra hardware we need is a strip of ten WS2812B LEDs attached to GPIO 2 that will sparkle when the wand is flicked. Visit the emlearn Getting Started page at **rpimag.co/emlearnpico** and follow the instructions for "ARM Cortex M4F/M33/M7 etc" to install it on your Raspberry Pi Pico. You'll also need to upload **led_helpers.py** from the GitHub repo (located in the **ch06** folder) to your Pico's **/lib** directory.

You can find the MicroPython code for this in **ch06-wand.py**:

```python
import emlearn_trees
import array
import gc
import _thread
from time import import sleep, ticks_ms
import board
import busio
import adafruit_lsm303_accel
import neopixel
from led_helpers import hsv_to_rgb, sparkle

multiplier = 1000
run_sparkle = False
np = neopixel.NeoPixel(machine.Pin(2), 10)

def core1_loop():
    global run_sparkle, np
```

```python
    while True:
        while not run_sparkle:
            sleep(0.1)
        sparkle(np,10,7,3,5,0.9)
        run_sparkle = False


_thread.start_new_thread(core1_loop, ())


i2c = busio.I2C(board.GP17, board.GP16)
accel = adafruit_lsm303_accel.LSM303_Accel(i2c)


print("loading model")
model = emlearn_trees.new(300, 10000, 200)
with open('flick_model.csv', 'r') as f:
    emlearn_trees.load_model(model, f)


resout = array.array('f',[0,0])
window = [0] * 30


print("running")
while True:
    del window[:3] # Remove the first 3 elements
    reading = accel.acceleration
    window.append(int(reading[0] * multiplier))
    window.append(int(reading[1] * multiplier))
    window.append(int(reading[2] * multiplier))
    model.predict(array.array('h', window), resout)
    if(resout[1] > 0.60):
        print(f"flick detected at {ticks_ms()} ",
            f"{resout[1]}% certainty")
        run_sparkle = True
        # Clear the window to avoid multiple detections
        window = [0] * 30
    sleep(0.1)
```

This code continuously runs two things. On core 0, it runs the model
every 0.1 seconds. If this model detects a flick with more than 0.6 confidence (on a scale of 0 to 1), it sets the **run_sparkle** global variable to **True**.
On core 1, it runs a loop that waits until **run_sparkle** is **True**, then runs
the **sparkle** function (imported from **led_helpers**). This function selects

a random set of LEDs and lights them up with a random colour, fading them in and out over a random time. To give them a consistent look, the saturation is set to 0.9, which gives them a pastel effect.

We won't go through the **led_helpers** code here, but it's straightforward and not the key focus of this project. It's in the GitHub repository (see *Example code* on page ix) if you'd like to take a closer look.

If you'd like the code to run automatically when your Pico 2 is running from an external power supply, rename the program to **main.py** (right-click **ch06-wand.py** in the file browser and click **Rename**).

Let's take a look at the hardware for making this work. You can use one or two LED strips for this (see **Figure 6-2**). Two will give your wand a more consistent light around the whole shaft. Since we want the same pattern displayed on both sides of the wand, we can simply split the data line so that both LED strips are driven by the same GPIO pin. This means that we don't have to worry about keeping them in sync in software.

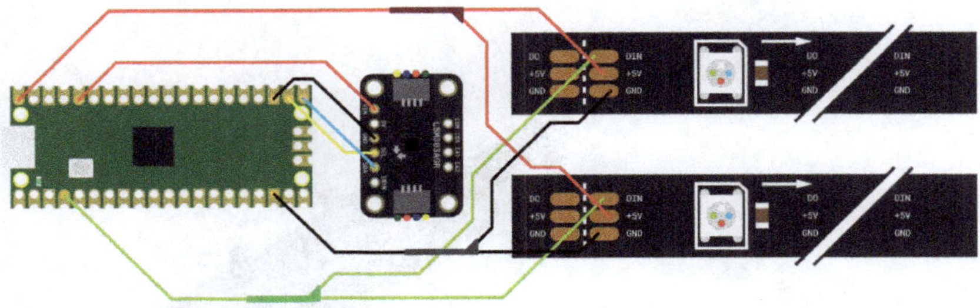

Figure 6-2   If using two strips, you can mount them back-to-back to provide 360 degrees of illumination to your wand

SOLDERING ONTO LED STRIPS

Most LED strips have periodic cut-points that allow you to get the length you want. On 5 V strips, these are typically every LED, but on higher voltage strips they can be less frequent. You should cut with sharp scissors, aiming to keep equal amounts of exposed copper on each side.

To solder onto these strips, first add a small amount of solder to the pad, then add a little solder to a wire, then you can join the two together with no additional solder.

That's the electronics sorted. Now we need a way of mounting everything. Obviously you can get as creative as you like with your enclosure. We've designed a basic 3D-printable frame that you can use to hold everything together. There are mounting holes that let you use 16 mm M2 nuts and bolts (you'll need eight in total), and you can cable-tie the WS2812B pixels in place (it's long enough for nine LEDs from a 60-LED-per-metre strip). The 3D print file is in the book's GitHub repository in the **ch06** directory (see *Example code* on page ix).

The wand takes power from USB because we've found USB battery packs to be the most versatile option for our projects. However, if you'd like to make it standalone, you can add an off-the-shelf battery module. A white plastic tube slotted over the LEDs will diffuse the glow and give a more magical effect.

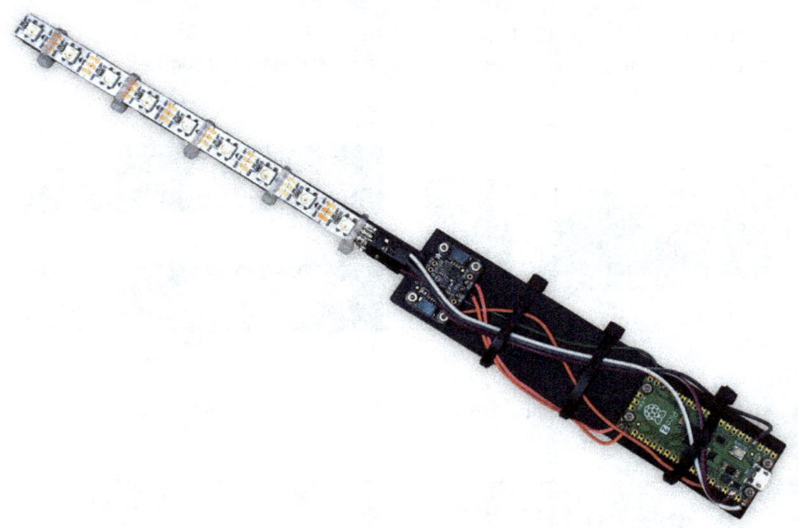

Figure 6-3   The electronic heart of a wand toy or costume prop — you can decorate it however you like

This first example is a full run-through of what it takes to build a machine learning setup from scratch. We gathered training data, picked a model, preprocessed the data, fed it into the model, then loaded the model onto our Pico 2, and finally ran it. For small tasks like our wand, this works well and can produce great results especially as the model is tailored for our particular sensor and setup.

However, as you take on more complex tasks, this approach gets harder. Gathering enough training data can be difficult (in general, the more complex the task, the more training data you need). Model selection, tuning, and preprocessing are all more critical and it can be harder to find a method that produces good results. On top of that, you may need expensive hardware to train your model on. That's the bad news. The good news is that once you have the model, the rest is easy.

Chapter 7

Speech to text and back again

Ears to hear and a voice to speak

Voice assistants such as Amazon Echo and Apple's Siri were among the first to make high-quality speech recognition widely available. Early iterations could interpret voice input without the tedious and repetitive training that previous generations of voice recognition software required. Although these voice assistants excelled at interpreting and generating speech, they lacked the charming conversational skills possessed by modern LLMs.

Even a dumb voice assistant needs to put in the hard work: modern voice recognition falls within the domain of artificial intelligence, relying on trained models and inference. There's nothing that says your voice assistant can't have a more powerful LLM as its engine. There's also nothing that says your voice assistant needs to be especially smart, either: perhaps you just need it to be able to recognize (and act on) certain action phrases, such as "turn on the lights" or "fire the nuclear weapons." You can do this with Raspberry Pi.

This chapter will look at speech recognition and text-to-speech separately and then will dive into a couple of applications of both technologies: live translation of speech, and a voice-powered chatbot. Most of the examples will run fine on a Raspberry Pi 5 without an AI HAT+, but the voice-powered chatbot will run an LLM on the AI HAT+ 2 while offloading the listening and speaking to the CPU.

WHAT YOU'LL NEED

- Raspberry Pi 5 8GB (or 16GB)
- Solderless breadboard
- 1 Push-button switch
- 1 LED
- 1 resistor (between 100 and 330 Ω)
- 3 socket-to-pin jumper wires
- 1 pin-to-pin jumper wire
- Microphone or audio adapter

Raspberry Pi 5 does not have a microphone jack. If you want to speak to your Raspberry Pi, you'll need a Bluetooth headset, USB microphone, or a USB headset. You could also use a USB audio adapter and plug a microphone into its audio input plug. This chapter will also show you how to capture system audio, so if you don't have a microphone handy, you can still follow along.

By far, the least troublesome configuration is to attach a Raspberry Pi-compatible USB audio adapter and connect a microphone and speaker to it. These adapters are often available from Raspberry Pi resellers.

Using a USB audio adapter helps for two reasons. First, Raspberry Pi OS should detect the device and set it as your audio input/output device automatically. Second, many sound cards include *Acoustic Echo Cancellation* (AEC), which automatically "subtracts" the audio coming out of the speaker from the microphone input. Later examples in the book require you to press a button before speaking (to avoid recording audio that comes out of your speaker).

If you are using the same sound device for input and output, you may find that you don't need to use the button, so you may be able to remove the **if** statement that returns from functions when they button is not pressed. You'll see functions like this in "Talking back (text-to-speech)" on page 147 and "Voice-powered chatbot" on page 154.

Speech recognition: Moonshine

Moonshine Voice is an open source collection of tools for adding real-time voice processing to your projects. It uses AI models that run on your Raspberry Pi, so you don't need to be connected to the internet. Moonshine is private, so your voice data never leaves your device. It also runs quite well on Raspberry Pi's CPU, so it won't be using the Raspberry Pi AI HAT+ accelerator.

To get started, open a Terminal or connect to your Raspberry Pi with SSH. Next, set up a virtual environment with the `--system-site-packages` option (so the environment can access modules installed with **apt**), activate it, and use **pip** to install `moonshine-voice`:

```
python3 -m venv ~/.virtualenvs/voice --system-site-packages
source ~/.virtualenvs/voice/bin/activate
pip install moonshine-voice
```

You'll be using this virtual environment for all the examples in this chapter. You can use whichever code editor you prefer to edit files.

Moonshine uses the Python `sounddevice` library, which is installed as one of Moonshine's dependencies automatically. The `sounddevice` library depends on the PortAudio library, so you'll need to install it with **apt**:

```
sudo apt install libportaudio2
```

Troubleshooting

Linux audio can be a little tricky at times, and you might encounter situations where your microphone isn't capturing audio or your speakers aren't playing any sounds. Here are a few things you can try if you have difficulties with audio:

- If you are using a separate sound device for input and output, such as a microphone connected to a USB audio adapter for recording and an HDMI monitor for playback, you may have some trouble getting audio to play. If this happens, try installing the apt packages mentioned in "Capturing system audio" on page 136.

- ▸ If your microphone is plugged into a USB audio adapter that has a microphone in port as well as a headphone or speaker out port, try plugging speakers or headphones into the audio out port. Your Raspberry Pi may be trying to send audio there.

- ▸ If you're using a USB microphone and you find that it isn't capturing audio, install the PulseAudio apt packages described in "Capturing system audio" on page 136. Then, while you're running a program that should be capturing sound from your microphone, click the Raspberry Pi menu, select **Sound & Video**, then choose **Volume Control**. Switch to the **Recording** tab and try different options from the drop-down menu next to the text that reads **ALSA Capture from**.

- ▸ If you receive a `sounddevice.PortAudioError` that indicates an invalid sample rate when you run one of the programs, you might not have the right input device selected. Right-click the microphone icon in your toolbar and select the microphone you want to use, reboot, and try again.

If you have multiple audio devices connected, you may need to tell Raspberry Pi OS which one to use. To set your default recording device:

1. Right-click the microphone icon in the taskbar, and choose the input device you want to use, such as a microphone connected to a USB audio adapter, or a Bluetooth headset.

2. For good measure, reboot after you've installed PortAudio and selected your microphone.

Capturing system audio

If you want to transcribe system audio (for example, a live news stream playing in your web browser), you'll need to install a few more packages. This can be useful if you don't have a microphone handy. Although Raspberry Pi OS Trixie uses PipeWire for audio by default, you can install just those PulseAudio bits needed to allow system outputs to be used as inputs to the programs in this chapter:

```
sudo apt install pavucontrol pulseaudio-utils pulseaudio
```

To capture system audio with the programs from this chapter, you'll need to start playing some audio (preferably spoken word, otherwise you won't have anything to transcribe). For example, you could open a web browser to your favourite news radio station and start playing the stream.

With the audio playing, do the following to confirm that everything is ready for you to try transcribing the audio:

1. Click the Raspberry Pi menu, choose **Sound & Video**, then click **Volume Control** to open the **pavucontrol** application.

2. Next, click the rightmost button in the tab menu at the top of the window until you land on the **Input Devices** tab. Click the **Show** menu as shown in **Figure 7-1**, and choose **Monitors**.

3. You should see your audio output port, such as **HDMI/Display-port**, as shown in **Figure 7-2**. The lower bar should be moving in sync with the audio you hear.

4. Close the **Volume Control** window.

Figure 7-1   Showing the audio monitors on the Input Devices tab

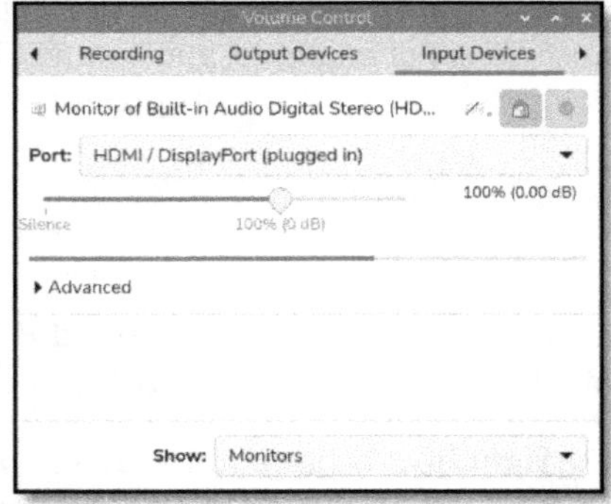

Figure 7-2   Examining the audio port

Now you can try the transcription example from the next section ("Python transcription" on page 139).

If you have more than one input device, you can switch the default device by running `pactl` at the command line. First, find a list of devices with the following command:

```
pactl list sources short
```

Here's some (abridged) example output on a Raspberry Pi with HDMI output and a Bluetooth headset connected:

```
58      alsa_output.platform-107c706400.hdmi.hdmi-stereo.monitor
256     bluez_output.38_F3_2E_B5_9D_3D.1.monitor
257     bluez_input.38:F3:2E:B5:9D:3D
```

To switch to using the Bluetooth headset as input, for example, you can run this command:

```
pactl set-default-source bluez_input.38:F3:2E:B5:9D:3D
```

Python transcription

Whether you're speaking into a microphone or transcribing system audio, you're now ready to run the transcription code. You can find this program in this book's GitHub repository (see *Welcome* on page v). Use `git clone` to check it out, then use the `cd` command to change to the **ai-projects-raspberrypi/ch07** subdirectory, as in:

```
cd ~/ai-projects-raspberrypi/ch07
```

Let's go through the source code of the transcription program, which you can find in the file named **transcribe_moonshine.py**. At the top of the program, we import some common Python modules (`time` and `sys`) as well as some key Moonshine modules and functions:

```
from moonshine_voice import (
    MicTranscriber,
    get_model_for_language,
    ModelArch,
    TranscriptEventListener,
)
import time
import sys
```

Here are brief descriptions of those Moonshine modules and functions:

- ▸ **MicTranscriber** — This is a class that will transcribe audio directly from a connected microphone (or system audio if you've configured your system to allow that).

- ▸ **get_model_for_language()** — This function looks up the Moonshine model based on the language and model type. You can find a list of available models at **rpimag.co/moonshine-models**.

- ▸ **ModelArch** — This enumeration contains the various model types: TINY, BASE, TINY_STREAMING, BASE_STREAMING, SMALL_STREAMING, and MEDIUM_STREAMING. TINY_STREAMING is a good choice for Raspberry Pi, because it does not use as many system resources as the larger models, but it is optimised for processing a stream of incoming text.

- ▸ **TranscriptEventListener** — This class contains helper functions that are called for various events that occur while Moonshine is performing a transcription.

Next, we define a class named `TextListener` that inherits from a parent class: `TranscriptEventListener`. There are several events that `TranscriptEventListener` can react to. By default, it does nothing with these events, so you need to override one or more of these methods to do anything with the transcription. Your event handler method will be called each time the event occurs.

The `on_line_started` event is called when transcription begins, and `on_line_completed` is called when it's finished.

Because Moonshine is designed for streaming input, it can update transcriptions on the fly. If you're writing an application that needs to show transcriptions as they happen, and update them when they change, the `on_line_updated` and `on_line_text_changed` events are helpful. The `on_error` event lets you do something when there is an error.

In this example, we'll only be using `on_line_completed`, and we simply print out the text of the transcription. We add the `flush` argument to print it immediately, rather than letting the output sit around in a buffer. This

will become important when we use *pipes* to send text from one Python
program to another.

```python
class TextListener(TranscriptEventListener):
    def on_line_completed(self, event):
        print(event.line.text, flush=True)
```

Next comes the main body of the code. We first load the tiny streaming
version of the English language model. Next, we set up a dictionary with
some options; it tells Moonshine to return the raw audio data, and to not
identify individual speakers.

The **vad_threshold** configures *Voice Activity Detection* (VAD) and can affect the length of transcribed segments (lower values tend to produce
longer segments). It then initialises the **MicTranscriber**, adds a **TextListener** to it, and starts the transcriber:

```python
# Load the model for the language we want to transcribe.
model_path, model_arch = get_model_for_language(
    "en", ModelArch.TINY_STREAMING
)


# Configure the transcriber.
options = {"return_audio_data": False,
           "identify_speakers": False,
           "vad_threshold": 0.2}
mic_transcriber = MicTranscriber(model_path=model_path,
                                 model_arch=model_arch,
                                 options=options)


# Add the listener to the transcriber and start it.
mic_transcriber.add_listener(TextListener())
mic_transcriber.start()
```

After the transcriber is up and running, the code runs an infinite loop,
briefly sleeping each time through, so as not to take up too much CPU
time. You can stop it by pressing **CTRL+C**, at which time, it will let you
know it's finished. Before the program exits, it stops and closes the transcriber to free any resources in use:

```python
print("CTRL+C to stop...", file=sys.stderr)
try:
    while True:
        time.sleep(0.1)
except KeyboardInterrupt:
    print("Finished.", file=sys.stderr)
finally:
    mic_transcriber.stop()
    mic_transcriber.close()
```

This code redirects its informational messages to standard error rather than the default (standard output), which means you can use a pipe (|) to redirect its output to another program while still seeing the informational messages (and without the informational messages being sent to that other program).

Run the program with **python transcribe_moonshine.py**. The first time you run it, it will download the speech recognition models (you can run offline after that). You'll probably see an error that says (among other things) "GPU device discovery failed" which you can safely ignore. When you see the message "CTRL+C to stop", start speaking into the microphone.

It will transcribe everything you say until you exit the program. If you've configured your Raspberry Pi to redirect audio output to an input, it will transcribe everything it captures from whatever audio you're playing. Run the following command:

```
$ python transcribe_moonshine.py
```

Here is some sample output:

```
CTRL+C to stop...
April is the cruelest month.
Breathing lilacs out of the dead land, mixing memory and
desire, stirring, dull roots with spring rain.
^CFinished.
```

It's not a perfect transcription, but it's not bad, especially for a tiny model!

If you see "input overflow" errors, it's the **sounddevice** library telling you that you're not reading data from PortAudio fast enough. You can try using the `MicTranscriber` constructor's `blocksize` argument to solve this:

```
mic_transcriber = MicTranscriber(model_path=model_path,
                                 model_arch=model_arch,
                                 blocksize=16384,
                                 options=options)
```

If that doesn't solve the problem, try installing the apt packages mentioned in "Capturing system audio" on page 136.

Python command recognition

Moonshine also includes support for command recognition. It will wait until it hears a predetermined phrase and will run a function of your choosing. Let's look at the command recognition example program, named **command_moonshine.py**. As with the transcription example, we start out by importing various modules and functions. In addition to the ones already described in "Python transcription" on page 139, you'll use:

- **IntentRecognizer** — This is a class that the transcriber will use to run different functions based on commands you speak. Moonshine refers to commands as *intents*. This class has a function, `register_intent()`, which you'll use associated phrases with functions that you define.

- **get_embedding_model()** — Similar to `get_model_for_language()`, this looks up a model based on its name and quantization type.

- **LED** — This class comes from the **gpiozero** library. It represents (and allows you to control) an LED connected to a GPIO pin.

As with the previous transcription example, this code imports several modules and functions from **moonshine_voice**. along with the **time** module and two classes from the GPIO Zero library. In place of `TranscriptEventListener`, the code imports the `IntentRecognizer` class and the `get_embedding_model()` function:

```python
from moonshine_voice import (
    MicTranscriber,
    get_model_for_language,
    ModelArch,
    IntentRecognizer,
    get_embedding_model,
)
import time
from gpiozero import LED, Button
```

> **EMBEDDING MODELS**
>
> An embedding model is trained on vast quantities of text, often in multiple languages. They represent an input (such as a sentence) as a mathematical vector. By comparing the similarity between two vectors, Moonshine can determine the similarity of two sentences. The **embeddinggemma-300m** model uses 768-dimensional vectors, and its 4-bit integers can represent 16 unique values, which means that there are 16^{768} possible combinations. This allows the embeddings to discern even subtle differences between sentences.

Next, the code initialises a few variables: one to control whether the code keeps running, the others to represent an LED and Button (we won't use the button until the next section). After that come the *intent functions* that the **IntentRecognizer** will invoke to turn words into actions. An intent function must accept three arguments, which will be passed into them each time the phrase is spoken: **trigger** (the phrase associated with the function), **utterance** (what the user actually said), **similarity** (a value between 0.0 and 1.0 indicating Moonshine's confidence level).

```python
running = True
button = Button(21)
led = LED(25)


def led_on(trigger: str, utterance: str, similarity: float):
    led.on()
    print("LED turned on.", flush=True)


def led_off(trigger: str, utterance: str, similarity: float):
```

```python
    led.off()
    print("LED turned off.", flush=True)

def quit(trigger: str, utterance: str, similarity: float):
    global running
    running = False
    print("I'm glad we had this little talk.", flush=True)
```

Next comes the main body of the code. It loads the embedding model and sets up the **IntentRecognizer**, At the time of this writing, the only supported model is the Gemma-based 300M parameter sentence embedding model (**embeddinggemma-300m**). A **q4** quantization (4-bit integer) reduces the original model size (32-bit floating point) down to something that performs well on a Raspberry Pi. After creating the **IntentRecognizer** as a variable named **recogniser**, the code registers the intent functions with it:

```python
# Load the embedding model for intent recognition.
embeddings_path, embeddings_arch = get_embedding_model(
    "embeddinggemma-300m", "q4"
)


# Set up the intent recognizer and register some intents.
recogniser = IntentRecognizer(
    model_path=embeddings_path, model_arch=embeddings_arch,
    model_variant="q4", threshold=0.6
)
recogniser.register_intent("turn on the light", led_on)
recogniser.register_intent("turn off the light", led_off)
recogniser.register_intent("quit", quit)
```

Finally, the code sets up the **MicTranscriber**. This code is very similar to the code from the previous example. The notable exception is that we add an intent recogniser object as the listener, rather than a **TranscriptEventListener**. After the transcriber is set up, the code goes into a loop that runs until **running** is set to False:

```python
# Configure the transcription engine.
model_path, model_arch = get_model_for_language(
    "en", ModelArch.TINY_STREAMING
)
```

```python
options = {"return_audio_data": False,
          "identify_speakers": False,
          "vad_threshold": 0.2}
mic_transcriber = MicTranscriber(model_path=model_path,
                                 model_arch=model_arch,
                                 options=options)

# Add the recognizer to the transcriber, and start it.
mic_transcriber.add_listener(recogniser)
mic_transcriber.start()

print('Say "quit" to stop...', flush=True)
while running:
    time.sleep(0.1)

mic_transcriber.stop()
mic_transcriber.close()
```

Before you run the program, make sure you shut down your Raspberry Pi, and disconnect it from power. Then, connect a momentary pushbutton and an LED with a current-limiting resistor of between 100 and 330 Ω as shown in **Figure 7-3**. You'll be using the button to let the program know when you're speaking. We're using the button so that the transcription engine doesn't accidentally start listening to itself when we add the text-to-speech engine.

Next, power up your Raspberry Pi and open a Terminal or SSH into it. If needed, activate your virtual environment and then run the example program with **python command_moonshine.py**. Try using the utterances to turn the LED on and off. When you've had enough fun, say "quit" to end the program.

You should see output like this:

```
Say "quit" to stop...
LED turned on.
LED turned off.
I'm glad we had this little talk.
```

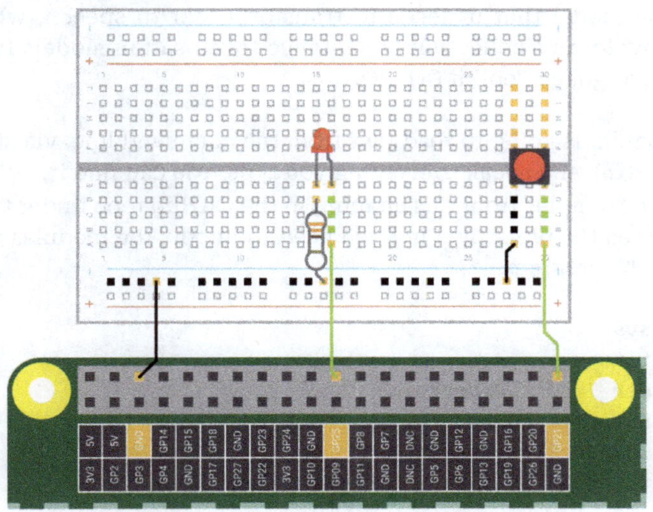

Figure 7-3 Connecting an LED to control with your voice

If you find that Moonshine doesn't understand something you're saying, try a synonym. For example, you can say "exit" instead of "quit," because it is similar enough to reach the threshold for a match. You can also try saying something like "switch the lamp on" rather than "turn on the light." Keep the button and LED wired up for now, because you'll need them in the next and final project.

Talking back (text-to-speech)

We're going to use the Piper TTS (text-to-speech) library to give a voice to your projects. You'll need the LED and button connected as described in "Python command recognition" on page 143. Make sure that you activated the virtual environment you used in previous examples and run these commands in the **ch07** directory:

```
pip install piper-tts sounddevice nltk
python3 -m piper.download_voices en_US-lessac-medium
python -c "import nltk; nltk.download('punkt_tab')"
```

The first command installs the **piper-tts** module. The second downloads the model that it needs to transform text to speech, while the third downloads a collection of sentence tokenisation models from the Natural Language Toolkit (NLTK).

This program is going to read text from the command line, via standard input (**stdin**), and speak everything you type. You can find it, along with the other examples from this chapter, in the GitHub repo under the **ch07** directory as **tts_piper.py**. First, we need to import the modules we'll be using in the program:

```python
import sys
import time
from piper import PiperVoice
import numpy as np
import sounddevice as sd
from nltk.tokenize import sent_tokenize
```

We're going to approach this a little differently than the previous examples, by first defining a class to make it easier to use Piper TTS from other programs. You'll use this class later when creating an interactive voice chatbot. The constructor takes the model filename as an argument and then loads the model. It configures an output stream, starts it, and waits half a second to give it time to warm up:

```python
class TTSApp:

    def __init__(self, model):
        self.voice = PiperVoice.load(model)
        sample_rate = self.voice.config.sample_rate
        self.stream = sd.OutputStream(samplerate=sample_rate,
                                      channels=1, dtype='int16')
        self.stream.start()
        time.sleep(0.5)  # Give the stream a moment to start
```

The remaining class methods are **speak()** and the destructor **__del__()**. The **speak()** method first breaks the text into separate sentences using NLTK's sentence tokeniser, then synthesises the spoken audio. If we didn't break the text up into separate sentences, we found that Piper TTS would only speak the last sentence of the text.

It then converts it into a form that can be played back and writes it to the audio stream. The destructor (**__del__**) is automatically called when an instance of the class goes out of scope, such as when the program is finished and is shutting down:

```python
def speak(self, text):
    for sentence in sent_tokenize(text):
        for chunk in self.voice.synthesize(sentence):
            data = np.frombuffer(chunk.audio_int16_bytes,
                                 dtype=np.int16)
            self.stream.write(data)

def __del__(self):
    self.stream.stop()
    self.stream.close()
```

The main loop of the program is contained under an **if __name__ == "__main__"** statement, which means it will only run if you are running this Python program directly from the command line. If you import it into another Python program, the main loop won't be called.

The main loop reads from the **sys.stdin** stream until there is nothing else to be read, sending each line of text to the **speak()** function:

```python
if __name__ == "__main__":
    app = TTSApp("./en_US-lessac-medium.onnx")

    print(f"Ready.", file=sys.stderr)
    for text in sys.stdin:
        app.speak(text.strip())

    print(f"Text-to-speech completed.", file=sys.stderr)
```

If you'd like to use a different language, you'll need to download the model for that language. The Piper project offers a web-based tool you can use to view and test different voices at **rpimag.co/pipersamples**. Once you find a voice you need, click the **Download** link. You can either download both the **onnx** and **json** files directly from there or pass the base name of the model (without the file extension) to the **piper.download_voices** command shown at the beginning of this section. Then you'll need to change the `PiperVoice.load()` command to load the corresponding **onnx** file.

Now you're ready to try it out. Run the command `python tts_piper.py` and try typing some text for it to speak. Press **CTRL+D** when you're done.

Next, let's give a voice to your command recognition program. Because we encapsulated the Piper functionality in a class, we need to only make some small changes to the **command_moonshine.py** program. You can find the completed version in the file **command_piper.py**. First, you need to add the following line at the top of your program somewhere:

```
from tts_piper import TTSApp
```

Next, add the following line right after the line that reads `led = LED(25)`:

```
tts_app = TTSApp("./en_US-lessac-medium.onnx")
def say(text):
    tts_app.speak(text)
```

Then, add the following two lines to the top of `led_on()`, `led_off()`, and `quit()`, which will cause them to ignore commands unless you keep your finger on the button:

```
    if not button.is_pressed:
        return
```

Finally, find every `print()` statement in the program, and change it to `say()`, removing the `, flush=True`. For example:

```
print("LED turned on.", flush=True)
```

would become:

```
say("LED turned on.")
```

Next, try running the program and giving it some commands:

```
python command_piper.py
```

Hold down the button when you wish to speak and release it a second or two after you are done speaking. Instead of printing the output to the command line, your Raspberry Pi should talk back to you! You'll want to keep the button and LED wired up for the final project in this chapter, so don't be in a hurry to disconnect them.

> **MOONSHINE OR STANDALONE PIPER?**
>
> Although Moonshine has built-in text-to-speech capabilities, we found that the Piper TTS module could synthesise speech faster. However, given Moonshine's rapid pace of development, we expect that their TTS engine will improve quickly, so we included a version of the earlier example ("Python command recognition" on page 143) that uses Moonshine's text-to-speech engine (in the repository as **ch07/command_moonshine_tts.py**). If you'd like to see how the earlier example was changed, run the following command:
>
> ```
> diff command_moonshine.py command_moonshine_tts.py
> ```

Machine translation

Marian NMT (Neural Machine Translation) is a very fast, efficient translation toolkit developed by researchers at Adam Mickiewicz University in Poznań and University of Edinburgh as well as Microsoft Research. The Helsinki-NLP (Natural Language Processing) Research Group at the University of Helsinki has trained a huge number of translation models using the Marian NMT framework and made them freely available as part of the Open Parallel Corpus Machine Translation (OPUS-MT) project.

The models are available in formats that are compatible with a variety of frameworks, but we'll use the Hugging Face *transformers* library that we introduced back in Chapter 4, *Generate images from prompts*. Before you install them, make sure you've activated your virtual environment (`source ~/.virtualenvs/voice/bin/activate`), and run the following commands:

```
pip install transformers nltk torch sentencepiece sacremoses
python -c "import nltk; nltk.download('punkt_tab')"
```

The first command installs the Python packages needed, and the second downloads a collection of sentence tokenization models from the Natural Language Toolkit (NLTK).

Now you're ready to write a little program to translate sentences. Because the program will need to download models from the Hugging Face Hub, download speed is limited by default. But you can speed things up by signing up for a Hugging Face account and setting an environment variable (**HF_TOKEN**) with your token string. See **rpimag.co/hf-tokens**.

It's easy to use these models. Let's go through the source of the example program, **translate_opus.py**. First, **import** the required libraries:

```python
from transformers import MarianMTModel, MarianTokenizer
import sys
from nltk.tokenize import sent_tokenize
import signal
```

Next, initialise the tokeniser and model:

```python
# Load the model and tokenizer.
mt_model_name = "Helsinki-NLP/opus-mt-en-de"
tokeniser = MarianTokenizer.from_pretrained(mt_model_name)
mt_model = MarianMTModel.from_pretrained(mt_model_name)
```

There are two types of tokenisers here. The **sent_tokenize()** function from the **nltk.tokenize** model takes care of breaking text up into sentences. The **MarianTokenizer** (**tokeniser**) converts each sentence into a tensor where each word is represented as a token. For example, "Hello World" might be represented as a tensor with the values **42174, 360, 0**, where 42174 is an integer that uniquely represents "Hello", 360 uniquely represents "World", and 0 indicates the end of the sentence.

The code is also using the **signal** library to ignore **CTRL+C** signals. We do this because we'll eventually use this program in a *pipeline*, with the transcription program sending lines of text to the translation program until it's done. When we press **CTRL+C**, we want the transcription program to handle that signal and send an end-of-file (EOF) character to the translation program. When running it interactively, you'll be able to exit the translation program with **CTRL+D**, but if it gets truly stuck somehow, you can send it to the background with **CTRL+Z** and you'll see a message like `[1]+  Stopped`. Next, use `kill %1` (replace **1** with whatever was in brackets before the `Stopped` message) to shut it down.

Once everything is set up, the code prints a "Ready" message to standard error (that way, you can safely redirect the output of the script to a file without cluttering it with messages). It then loops continuously until it gets an EOF via standard input. For each bit of text it receives, the code splits it into separate sentences, tokenises them, and translates them. Once that's done, it prints out each sentence in the response:

```python
signal.signal(signal.SIGINT, signal.SIG_IGN)
print(f"Ready.", file=sys.stderr)
for text in sys.stdin:
    # Break the text into sentences.
    for sentence in sent_tokenize(text):

        inputs = tokeniser(sentence, return_tensors="pt")
        translated = mt_model.generate(**inputs)

        # Decode and print the translated sentences.
        decoded = tokeniser.decode(translated,
                                   skip_special_tokens=True)
        for sentence in decoded:
            print(sentence)
print("Translation finished.", file=sys.stderr)
```

You can test it out with `python translate_opus.py` (if needed, activate your virtual environment first). After it starts up and displays the "Ready" message, start typing English text into it, and watch it quickly translate it into German. The first time you run it, the informational messages about the model download may overlap with the "Ready" message. When you're done, press **CTRL+D** to send an EOF character, and it will exit.

You can now try sending the output of another program to the translation program. For example, to transcribe anything you say into the microphone, and translate it to German, you could try the following:

```
python transcribe_moonshine.py | python translate_opus.py
```

Speak a few words to it and you should see some output like this:

```
CTRL+C to stop...
Loading weights: 100%|███| 258/258 [00:00<00:00, 27062.06it/s]
Ready.
Guten Tag.
Hallo, Computer.
```

Press **CTRL+C** when you're done and the two programs should shut down gracefully. There are many more language models available from the Opus-MT project. The list is quite massive, but if you sort it by number of downloads, the most popular ones will rise to the top: **rpimag.co/ OpusMT-models**. The naming convention for the Opus-MT models that offer good results with minimal CPU and RAM impact is `Helsinki-NLP/ opus-mt-`*`from`*`-`*`to`* where you should replace `from` with the two-letter language code of the source language and `to` with the code for the target language. For example, to translate from Spanish to English, you'd want `Helsinki-NLP/opus-mt-es-en`.

Voice-powered chatbot

Let's build something that ties together transcription and text-to-speech, by creating a voice-powered chatbot that you can have entertaining conversations with. You'll need the LED and button connected as described in "Python command recognition" on page 143. You'll also need a Raspberry Pi AI HAT+ 2 because we're going to use its Hailo 10H AI accelerator to run a Large Language Model (LLM). Even with the 8 GB of dedicated AI accelerator memory, the available LLMs are quite small, and are more prone to hallucinations than online LLMs you may be familiar with.

You should first make sure your AI HAT+ 2 is installed as described in the online documentation at **rpimag.co/aihatinstall**. You will also need to complete the software setup (see "AI HAT+ 2 software prerequisites" on page 101). When you're finished, run this command and confirm that the Device Architecture is listed as HAILO10H:

```
hailortcli fw-control identify
```

Create a text-only chatbot

First, we'll build a Python module that can power a text-based chatbot, and then we'll use it to build a voice-powered one. When you installed the Hailo software, it included everything you need to load and run an LLM on your AI HAT+ 2. You just need to download a model and write a little code to interact with it.

You can find all of Hailo's generative models at **rpimag.co/hailo-gen-ai**. On that page, change the **All Products** filter to **Hailo-10H** and the **All Models** filter to **LLM**. We suggest starting out with the Qwen2-1.5B-Instruct model. Click the link for it and scroll down to the Technical Details section of the page. Click the link next to the text that reads Compiled Model to download it to your computer. Move the downloaded file (**Qwen2-1.5B-Instruct.hef**) to the **ch07** subdirectory of the example code you checked out from GitHub.

Now you're ready to build the LLM module and run a simple chatbot. The complete code for this is in **llm.py**. You probably won't be surprised to learn that the code starts out by importing all the modules we'll need:

```python
from hailo_platform import VDevice
from hailo_platform.genai import LLM
from rich.console import Console
import sys
import signal
```

You might notice that we didn't instruct you to install any `hailo` modules with pip. That's because you installed the Hailo modules when you followed the steps to set up the AI HAT+ 2. And you may recall that we asked you to use the `--system-site-packages` flag when you set up the environment at the beginning of this chapter. That flag ensures that your Python environment can locate those system modules.

Next comes the `LLMApp` class. It begins with the constructor, which takes three arguments: the path to the model, a console from the rich module, and the *system prompt* used to initialise the chat. The system prompt is something that sets the tone of the conversation and is usually something like "you are a helpful assistant."

It then creates a **VDevice** object to represent the Hailo hardware, then creates an **LLM** object (loading the model as it does so), and initialises the model with the system prompt:

```python
class LLMApp:

    def __init__(self, model_path, console, prompt):
        self.console = console

        self.console.print("Initialising device...")
        params = VDevice.create_params()
        params.group_id = "SHARED"
        self.vdevice = VDevice(params)

        self.console.print(f"Loading model {model_path}...")
        self.llm = LLM(self.vdevice, model_path)

        # Add system prompt to the LLM's context.
        self.console.print("Initialising model...")
        self.sysmsg = {"role": "system", "content": prompt}
        self.initialise_context()
```

The **initialise_context()** function clears any of the model's context and sends the system prompt to the model. We don't care about the model's response to the system prompt, so we only ask it for one token:

```python
    def initialise_context(self):
        self.llm.clear_context()
        # Only a single token is needed to add to the context.
        for token in self.llm.generate(prompt=[self.sysmsg],
                                       max_generated_tokens=1):
            pass
```

Next comes the heart of this class, the **generate()** function. It first checks whether the model's *context* is close to being exhausted. The context determines how much the model can remember of your chat. If you have a long-running conversation, you might run out of context, so when it gets close to being full, this function calls **initialise_context()** to start fresh. Unfortunately, this means that the LLM won't remember anything that came before.

Hailo's LLM object will maintain the conversation context between calls, so you do not need to manually manage the context by accumulating past interactions and re-sending them with each new message.

The **generate()** function then constructs a user message and asks the model to generate a response. As the response is generated, we use the console to show an animated spinner to let you know the model is working. There are two arguments we're using to condition the model before it responds. The temperature controls how much restraint the model should use. In general, a lower temperature gives you shorter, more factual responses. The maximum generated tokens value determines, to some extent, the length of the responses. If you have a high temperature and a low maximum number of tokens, the responses may be truncated.

Once it has the response, this function removes the end of message tag and returns it. One last function appears after **generate()**, the class destructor. It cleans up system resources that were in use.

```python
def generate(self, user_input,
            max_tokens=200, temperature=0.2):
    # Clear context if we're getting close to the limit.
    capacity = self.llm.max_context_capacity()
    usage = self.llm.get_context_usage_size()
    if usage / capacity > 0.90:
        self.console.print("Context full, clearing.")
        self.initialise_context()

    msg = {"role": "user", "content": user_input}
    r = ""
    try:
        with self.console.status("[blue]Generating..."):
            for token in self.llm.generate(
                        prompt=[msg],
                        temperature=temperature,
                        max_generated_tokens=max_tokens):
                r += token

    except Exception as e:
        self.console.log(f"Error occurred: {repr(e)}")
        sys.exit(1)
```

```python
        # Remove end-of-message tokens.
        return r.replace("&lt;|im_end|&gt;", "")

    def __del__(self):
        self.llm.clear_context()
        self.llm.release()
        self.vdevice.release()
```

Finally, the code defines a short demo program to run if you run this program directly. It sets up a console for diagnostics and responses, creates an instance of the **LLMApp**, and generates responses for everything you type until you press **CTRL+D** to signify the end of the stream.

```python
if __name__ == "__main__":
    diags = Console(stderr=True, style="purple")
    output = Console(style="dark_cyan")

    app = LLMApp("./Qwen2-1.5B-Instruct.hef", diags,
                 "You are a helpful assistant.")

    diags.print("Ready.")
    for text in sys.stdin:
        response = app.generate(text.strip())
        output.print(response)

    diags.print("Farewell from llm.py")
```

Try running it with **python llm.py**. Wait for it to display the text "Ready" and then try having a conversation with it. Press **CTRL+D** when you're done, and it will display "Farewell from llm.py" before shutting down.

```
$ python llm.py
Initialising device...
Loading model ./Qwen2-1.5B-Instruct.hef...
Initialising model...
Ready.
What is Raspberry Pi?

Raspberry Pi is a small computer that uses an open-source
```

operating system called Linux. It's often used for DIY
projects and as a platform to run software applications,
such as video games or programming languages like Python or
Ruby on Rails. The Raspberry Pi is also popular in educational
settings as a tool for teaching basic computer science
concepts and coding skills.

Add voice recognition and text to speech

With all the building blocks in place, it's time to write a voice-powered
chatbot. This will be one of the shorter programs because you're able to
reuse the **LLMApp** and **TTSApp** classes to build this. You can find this program in the **ch07** directory as **chatbot.py**. It starts out with the **import**
statements needed for all the module the program uses:

```python
from moonshine_voice import (
    MicTranscriber,
    get_model_for_language,
    ModelArch,
    TranscriptEventListener,
)
import time
from llm import LLMApp
from tts_piper import TTSApp
from rich.console import Console
from gpiozero import LED, Button
```

Next, it creates an LED and Button, then initialises the LLM and TTS apps:

```python
button = Button(21)
led = LED(25)


diags = Console(stderr=True, style="purple")
llm = LLMApp("./Qwen2-1.5B-Instruct.hef", diags,
            "You are a helpful assistant.")
tts_app = TTSApp("./en_US-lessac-medium.onnx")
```

After those are set up, the program defines the **TextListener** class that's
based on **TranscriptEventListener**. This is like the **TextListener** from

the transcription program you saw earlier but has slightly different be-
haviour. When it receives a complete line of text, it turns the LED on to
let you know that it's time to release the button. Then, it generates a response and uses the text-to-speech object to speak it out loud. Once it's
done, it turns the LED off to let you know it's finished.

Notice that it only performs the preceding steps if the button is held
down. If it's not, it simply displays the line of text that it's ignoring, which
helps you know when it's listening or not.

```python
class TextListener(TranscriptEventListener):
    def on_line_completed(self, event):
        if not button.is_pressed:
            diags.print(f"Skipping: {event.line.text}")
            return

        led.on()
        diags.print(f"Transcribed: {event.line.text}")
        response = llm.generate(event.line.text)
        tts_app.speak(response)
        led.off()
```

The rest of the program looks a lot like the main body of the **tran-
scribe_moonshine.py** program you saw earlier. It loads the Moonshine
model, adds a **MicTranscriber**, and adds the listener to the transcriber.
Then it goes into a loop and keeps running until you press **CTRL+C**. When
that happens, it cleans up the transcriber, then exits.

```python
# Load the model and set up the transcriber.
model_path, model_arch = get_model_for_language(
    "en", ModelArch.TINY_STREAMING
)
options = {"return_audio_data": False,
           "identify_speakers": False,
           "vad_threshold": 0.2}
mic_transcriber = MicTranscriber(model_path=model_path,
                                 model_arch=model_arch,
                                 options=options)
mic_transcriber.add_listener(TextListener())
mic_transcriber.start()
```

```python
diags.print("Ready! CTRL+C to stop...")
try:
    while True:
        time.sleep(0.1)
except KeyboardInterrupt:
    diags.print("Finished.")
finally:
    mic_transcriber.stop()
    mic_transcriber.close()
```

Now you can run this program with python chatbot.py. Wait for the text
"Ready! CTRL+C to stop..." to appear. When you're ready to speak to the
chatbot, press the button and hold it down, then try saying something.
Wait for the LED to come on, then release the button.

Next steps

We've left you a lot of room to customise these programs and experiment
on your own. For example, you might want to try adjusting the temperature and maximum number of tokens to tune the type of response you
get. Play around with the system prompt as well, and you can figure out
how to create specialised assistants.

You may also want to try turning the chatbot into a more useful assistant.
One thing you could try would be adding in an IntentRecognizer to let
you quit the program with a specific keyword. Intent recognisers and
transcript event listeners can happily coexist in the same program.

You can also take things a bit further and create an agent that's capable
of more than conversation and handling predefined actions. Hailo has
published a GitHub repository of apps that show off the full range of
capabilities and is frequently updated. The Interactive Chat Agent example (**rpimag.co/hailo-agent-demo**) uses a model that can call functions
based on your conversations with it.

Chapter 8

Recognise video and image content

Use the AI Camera or an AI HAT+ to recognise and classify objects in near real-time

Image classification and object detection are some of the most practical and immediate uses for Raspberry Pi hardware in the AI space. With object detection a model takes the input in the form of an image and locates and labels items inside the image. For example, "cat," "bus," or aeroplane". It does this using a convolutional neural network (CNN) that has been trained to extract abstract features from pixel data and map those features to class probabilities.

> **WHAT IS A CNN?**
>
> A convolutional neural network (CNN) is designed to process grid-like data (typically images). The central idea is that rather than examining each pixel independently, a CNN exploits that nearby pixels are related and is trained to detect patterns: edges, textures, and shapes that appear anywhere in the image. Once trained it can then detect those patterns in a grid.
>
> Stanford's CS231n is regarded as one of the best resources for understanding CNNs in detail: **rpimag.co/cs231n**.

You can perform object detection on a still image supplied as a file, or via a video file such as a YouTube stream. Importantly, for Raspberry Pi, you can also perform object detection on live video stream using an attached camera module.

Most Raspberry Pi boards have at least one Camera Serial Interface (CSI) connectors, and Raspberry Pi 5 has two. These enable you to connect a compatible camera directly to Raspberry Pi hardware. With GPIO hardware you can also attach actuators such as motors, solenoids, and relays to perform a real world response to object detection.

As outlined in Chapter 2, *The Raspberry Pi AI stack*, there are two main approaches to adding inferencing and performing object detection on Raspberry Pi:

AI Camera

> A camera module with an integrated Sony IMX500 sensor, which includes its own *neural processing unit* (NPU). This is best for specific vision tasks such as classification, detection, and pose estimation.

AI HAT+/AIHAT+ 2

> A HAT+ add-on board with a Hailo-8 or Hailo-10 (NPU). This is more general-purpose device that can process data from images, video files, and any type of attached camera. It has higher throughput than AI Camera for multiple concurrent models.

We'll look at both hardware approaches in this chapter.

Using Raspberry Pi AI Camera

Raspberry Pi AI Camera (see **Figure 8-1** and **Figure 8-2**) uses the Sony IMX500 imaging sensor to provide low-latency and high-performance AI capabilities to any camera application. Tight integration with Raspberry Pi's camera software stack (**rpimag.co/camerasoftware**) allows users to deploy their own neural network models with minimal effort.

This chapter demonstrates how to run either a pre-packaged or custom neural network model on AI Camera. Additionally, this section includes the steps required to interpret inference data generated by neural networks running on the IMX500 in **rpicam-apps** (rpimag.co/rpicamapps) and Picamera2 (**rpimag.co/picamera2**).

Figure 8-1   AI Camera connected to Raspberry Pi 5 using a ribbon cable

Let's see how to run the pre-packaged MobileNet SSD and PoseNet neural network models on the Raspberry Pi AI Camera.

These instructions assume you are using the AI Camera attached to either a Raspberry Pi 4 Model B or Raspberry Pi 5 board. With minor changes, you can follow these instructions on other Raspberry Pi models with a camera connector, including the Raspberry Pi Zero 2 W and Raspberry Pi 3 Model B+.

Figure 8-2   Camera connected to Raspberry Pi 5 with AI HAT+ connect to PCIe connector

Prerequisites

First, ensure that your Raspberry Pi has the latest software. Run the following command to update:

```
sudo apt update && sudo apt full-upgrade
```

Install the IMX500 firmware

The AI camera must download runtime firmware onto the IMX500 sensor during startup. To install these firmware files onto your Raspberry Pi, run the following command:

```
sudo apt install imx500-all
```

This command:

- installs the **imx500_loader.fpk** and **imx500_firmware.fpk** files into **/usr/lib/firmware**. These are required to operate the IMX500 sensor.

- places a number of neural network model firmware files in **/usr/share/imx500-models/**.

- installs the IMX500 post-processing software stages for use by the various *rpicam-apps*.

- installs the Sony network model packaging tools.

> **FIRMWARE LOADING TIMES**
>
> The IMX500 kernel device driver loads all the firmware files when the camera starts. This may take several minutes if the neural network model firmware has not been previously cached. The demos below display a progress bar on the console to indicate firmware loading progress.

Reboot and test

Now that you've installed the prerequisites, restart your Raspberry Pi:

```
sudo reboot
```

Test that the AI Camera is working with:

```
rpicam-hello
```

Run example applications

Once all the system packages are updated and firmware files installed, you can start running some example applications. As mentioned, the Raspberry Pi AI Camera integrates fully with libcamera, rpicam-apps, and Picamera2. These applications include:

- `rpicam-hello`: A "hello world"-equivalent for cameras, which starts a camera preview stream and displays it on the screen.

- **rpicam-jpeg**: Runs a preview window, then captures high-resolution still images.

- **rpicam-still**: Emulates many of the features of the original **raspistill** application.

- **rpicam-vid**: Captures video.

- **rpicam-raw**: Captures raw (unprocessed Bayer) frames directly from the sensor.

- **rpicam-detect**: Not built by default, but users can build it if they have TensorFlow Lite installed on their Raspberry Pi. Captures JPEG images when certain objects are detected. See **rpimag.co/rpicamdetect** for more details.

Raspberry Pi OS includes the five basic *rpicam-apps*, so you can record images and videos using a camera with a fresh install of Raspberry Pi OS.

You can create your own rpicam-based applications with custom functionality to suit your own requirements. The rpicam-apps source code is freely available under a BSD-2 Clause licence (**rpimag.co/rpicamapps**).

Adding on-sensor AI

There are two detection models installed alongside the AI Camera. These can be found as *json* models (alongside others installed during setup) with:

```
ls /usr/share/rpi-camera-assets/
```

Use **less** to view the contents of a file (press **Q** to exit):

```
less /usr/share/rpi-camera-assets/imx500_mobilenet_ssd.json
```

The two models of interest to us are:

- MobileNet SSD object detection directly on the IMX500's NPU. Detects and classifies objects (people, cars, animals, etc.) and outputs bounding box metadata.

- **imx500_posenet.json**: Runs PoseNet on the IMX500's NPU to detect human body keypoints (shoulders, elbows, knees, etc.) — i.e. skeleton/pose estimation.

You can find the two models they refer to by looking in the directory **/usr/share/imx500-models/**:

- **imx500_network_ssd_mobilenetv2_fpnlite_320x320_pp.rpk**

- **imx500_network_posenet.rpk**

Adding object detection

The MobileNet SSD neural network performs basic object detection, providing bounding boxes and confidence values for each object found.

You can locate the **imx500_mobilenet_ssd.json** file in the directory **/usr/share/rpi-camera-assets/**, contains the configuration parameters for the IMX500 object detection post-processing stage using the MobileNet SSD neural network.

imx500_mobilenet_ssd.json declares a post-processing pipeline that contains two stages:

`imx500_object_detection`

This stage picks out bounding boxes and confidence values generated by the neural network in the output tensor.

`object_detect_draw_cv`

This stage draws bounding boxes and labels on the image.

The MobileNet SSD tensor requires no significant post-processing on your Raspberry Pi to generate the final output of bounding boxes. All object detection runs directly on the AI Camera.

Attach the *json* file to the AI Camera command using the `--post-process` option. You will also need to use other commands to set the timeout to zero (or the preview will cancel immediately). You can also set options for determining the viewfinder width and height, and the framerate.

The following command runs **rpicam-hello** with object detection post-processing enabled:

```
rpicam-hello -t 0s --post-process-file \
  /usr/share/rpi-camera-assets/imx500_mobilenet_ssd.json \
  --viewfinder-width 1920 --viewfinder-height 1080 \
  --framerate 30
```

TURN rpicam-hello INTO A MIRROR

If you'd prefer to make the **rpicam-hello** preview window behave like a mirror, you can add the **--hflip** argument to the command line. To get more information on the available commands, run **rpicam-hello --help** to view their options. This will make the pose estimation demos feel a bit more natural.

After running the command, you should see a viewfinder that overlays bounding boxes on objects recognised by the neural network (see **Figure 8-3**). Note it detects two objects in our test scene from a maximum of five objects (this limit is determined by the post processor file's **max_detections** line). To exit, switch back to the Terminal you used to launch **rpicam-vid** and press **CTRL+C**.

Figure 8-3   Object detection with AI Camera.

To record video with object detection overlays, use **rpicam-vid** instead.
The following command captures a video with object detection post-processing enabled:

```
rpicam-vid -t 10s -o output.264 --post-process-file \
  /usr/share/rpi-camera-assets/imx500_mobilenet_ssd.json \
  --width 1920 --height 1080 --framerate 30
```

In the **rpicam-hello** example before that one, we used **--viewfinder-
width** and **--viewfinder-height** to control the size of the preview window. The **--width** and **--height** arguments control the resolution of the
output file rather than of the preview window.

The **output.264** file contains raw video, which may be useful to some, but
is difficult to view easily. Choose a different format such as mp4 if you
want to view the video file. Name the output file **output.mp4** and *rpicam-
apps* will select the right container automatically.

```
rpicam-vid -t 10s -o output.mp4 --post-process-file \
  /usr/share/rpi-camera-assets/imx500_mobilenet_ssd.json \
  --width 1920 --height 1080 --framerate 30
```

Now you can view the file with a media application such as VLC:

```
vlc output.mp4
```

Post processor file

View the available post processor files with:

```
ls /usr/share/rpi-camera-assets/
```

You'll find a selection of *json* files that can be passed into Raspberry Pi
Camera modules as a post processor file:

- **acoustic_focus.json**

- **annotate_cv.json**

- ▸ **face_detect_cv.json**

- ▸ **hdr.json**

- ▸ **imx500_mobilenet_ssd.json**

- ▸ **imx500_posenet.json**

- ▸ **motion_detect.json**

- ▸ **negate.json**

- ▸ **sobel_cv.json**

Only two of these require the AI Camera and both are prefixed with imx500: **imx500_mobilenet_ssd.json** and **imx500_posenet.json**. You can view the example *json* files with a pager such as `less` as shown in the previous section.

You can configure the **imx500_object_detection** stage in many ways. First, make a copy of the file and then edit it with a text editor. You should create a directory for your custom files before you copy it:

```
mkdir ~/rpi-camera-assets-custom
cd ~/rpi-camera-assets-custom
cp /usr/share/rpi-camera-assets/imx500_mobilenet_ssd.json .
nano imx500_mobilenet_ssd.json
```

For example, `max_detections` defines the maximum number of objects that the pipeline will detect at any given time. `threshold` defines the minimum confidence value required for the pipeline to consider any input as an object.

The raw inference output data of this network can be quite noisy, so this stage also performs some temporal filtering and applies hysteresis. To disable this filtering, remove the `temporal_filter` config block. To use a customised version of the json file, you can use that filename with the `--post-process-file` argument, as in:

```
--post-process-file \
    ~/rpi-camera-assets-custom/imx500_mobilenet_ssd.json
```

Pose estimation

The PoseNet neural network performs pose estimation, labelling key points on the body associated with joints and limbs. **imx500_posenet.json** contains the configuration parameters for the IMX500 pose estimation post-processing stage using the PoseNet neural network model.

imx500_posenet.json declares a post-processing pipeline that contains two stages:

`imx500_posenet`

> This stage fetches the raw output tensor from the PoseNet neural network model

`plot_pose_cv`

> This stage draws line overlays on the image

The AI Camera performs basic detection, but the output tensor requires additional post-processing on your host Raspberry Pi to produce the final output.

The following command runs **rpicam-hello** with pose estimation post-processing (**Figure 8-4**):

```
rpicam-hello -t 0s --post-process-file \
  /usr/share/rpi-camera-assets/imx500_posenet.json \
  --viewfinder-width 1920 --viewfinder-height 1080
```

Figure 8-4 Pose detection

You can configure the **imx500_posenet** stage in many ways. For example, **max_detections** defines the maximum number of bodies that the pipeline will detect at any given time. **threshold** defines the minimum confidence value required for the pipeline to consider input as a body.

Accessing the AI Camera from Python

Picamera2 is a Python library that provides convenient access to the camera system of Raspberry Pi. Picamera2 is built on top of the Python bindings provided by the open source libcamera project.

For examples of image classification, object detection, object segmentation, and pose estimation using Picamera2, see the Picamera2 GitHub repository (**rpimag.co/picamera2**) as well as the Picamera2 manual (**rpimag.co/picamera2manual**)

Most of the examples use OpenCV for some additional processing. To install the dependencies required to run OpenCV, run this command:

```
sudo apt install python3-opencv python3-munkres
```

Now download the picamera2 repository to your Raspberry Pi to run the examples. You'll find example files in the root directory, with additional information in the **README.md** file.

```
cd ~
git clone https://github.com/raspberrypi/picamera2.git
cd picamera2/examples/imx500
```

Run the following script from the repository to run object detection:

```
python imx500_object_detection_demo.py
```

Although you can close the Python preview window, this won't quit the application. You'll still need to press **CTRL+C** in the Terminal window you used to run the program.

> ### MAKE IT A MIRROR
>
> If you'd prefer that the preview window behave like a mirror, you'll need to flip it horizontally. Unlike `rpicam-hello`, it's not as simple as adding the `--hflip` argument. You'll need to edit the Python file and make two changes:
>
> - Add the line `from libcamera import Transform` to the top of the file, along with the other `import` statements.
>
> - Find the line that invokes the `create_preview_configuration()` function, add a comma at the end of the existing arguments, then add the argument `transform=Transform(hflip=1)`.

You can pass a specific model to the Python code (note that there must be no space before the \ on the second line, and no space before **fpnlite_** on the third line):

```
python imx500_object_detection_demo.py --model \
  /usr/share/imx500-models/imx500_network_ssd_mobilenetv2_\
fpnlite_320x320_pp.rpk
```

> **PICK A MODEL**
>
> The imx500 demo expects post processing. Valid models from the **imx500-models** folder end with pp.

To try pose estimation in Picamera2, run the following script from the Picamera2 repository:

```
python imx500_pose_estimation_higherhrnet_demo.py
```

Under the Hood

The Raspberry Pi AI Camera works differently from traditional AI-based camera image processing systems, as shown in **Figure 8-5**.

The left side demonstrates the architecture of a traditional AI camera system. In such a system, the camera delivers images to the Raspberry Pi. The Raspberry Pi processes the images and then performs AI inference. Traditional systems may use external AI accelerators (as shown) or rely exclusively on the CPU.

The right side demonstrates the architecture of a system that uses IMX500. The camera module contains a small Image Signal Processor (ISP) which turns the raw camera image data into an input tensor. The camera module sends this tensor directly into the AI accelerator (NPU) within the camera, which produces an output tensor that contains the inferencing results. The AI accelerator sends this tensor to the Raspberry Pi. There is no need for an external accelerator, nor for the Raspberry Pi to run neural network software on the CPU.

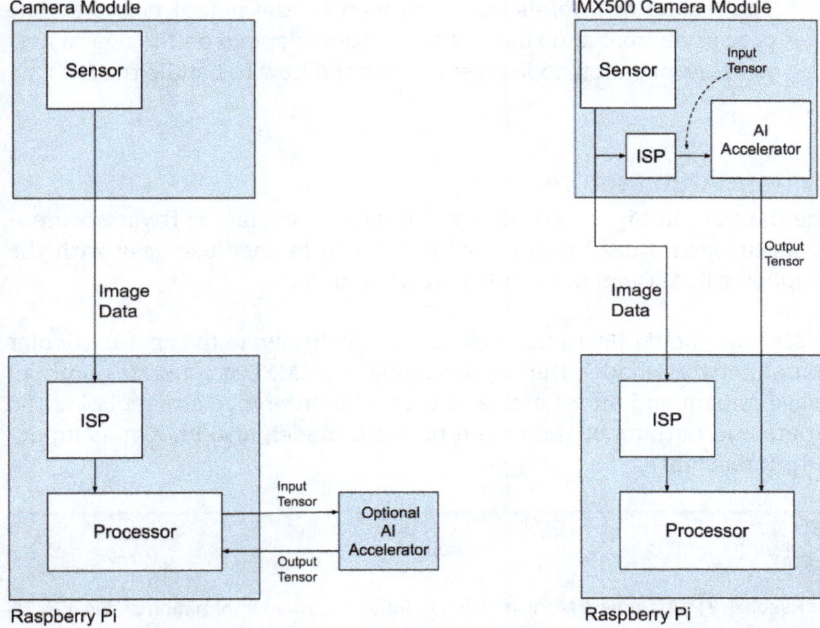

Figure 8-5   Camera Module vs AI Camera Module

To fully understand this system, familiarise yourself with the following:

Input Tensor

> The part of the sensor image passed to the AI engine for inferencing. Produced by a small on-board ISP which also crops and scales the camera image to the dimensions expected by the neural network that has been loaded. The input tensor is not normally made available to applications, though it is possible to access it for debugging.

Region of Interest (ROI)

> Specifies exactly which part of the sensor image is cropped out before being rescaled to the size demanded by the neural network. Can be queried and set by an application. The units used are always pixels in the full resolution sensor output. The default ROI setting uses the full image received from the sensor, cropping no data.

Output Tensor

> The results of inferencing performed by the neural network. The precise number and shape of the outputs depend on the neural network. Application code must understand how to handle the tensor.

System Architecture

The diagram in **Figure 8-6** shows the various camera software components (in green) used during our imaging/inference use case with the Raspberry Pi AI Camera module hardware (in red).

At startup, the IMX500 sensor module loads firmware to run a particular neural network model. During streaming, the IMX500 generates both an image stream and an inference stream. This inference stream holds the inputs and outputs of the neural network model, also known as input/output tensors.

DEVICE DRIVERS

Raspberry Pi AI Camera Module documentation has more information on the device drivers used by AI Camera: **rpimag.co/aicamdrivers**

Set up the AI HAT+ or AI Kit

You can use AI HAT+ and AI HAT+ 2 to perform image inferencing with a regular camera. Both AI HAT+ models feature a Hailo NPU (Neural Processing Unit). The Hailo NPU is an AI accelerator chip designed to run neural networks; instead of Raspberry Pi's CPU doing the AI work, the NPU handles it more efficiently.

AI HAT+ hardware requires a Raspberry Pi 5 because it connects via the PCIe connector (which is only present on Raspberry Pi 5).

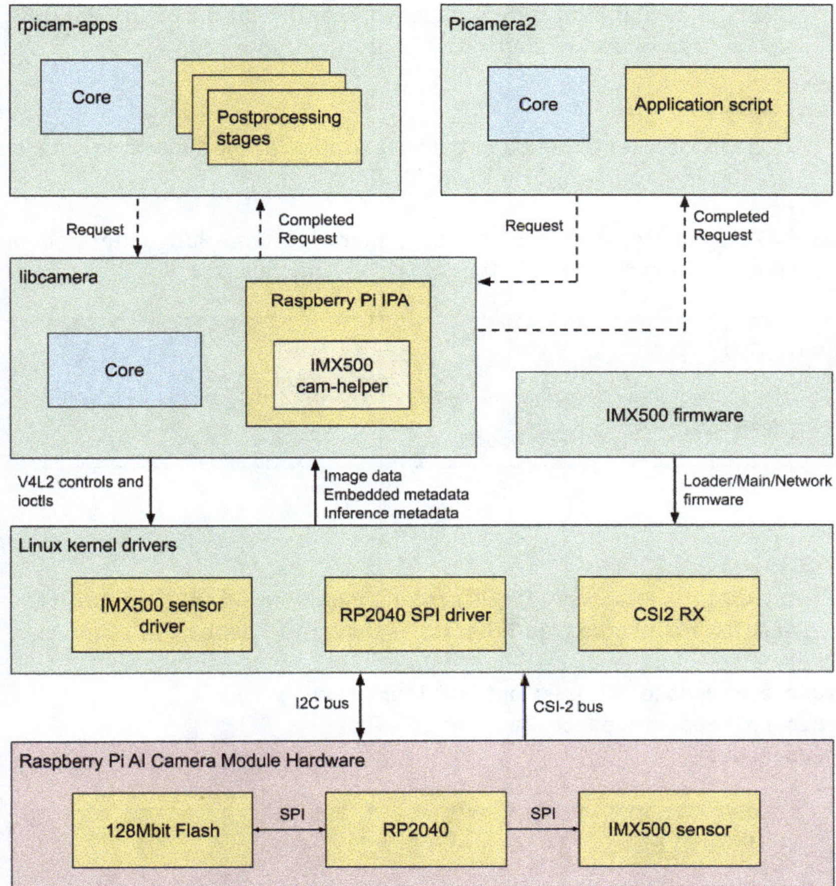

Figure 8-6 The various software components (green) and Raspberry Pi hardware components (red)

There are three Hailo powered options available:

Raspberry Pi AI Kit

This consists of an M.2 HAT+ with a pre-installed Hailo-8L NPU. Note this is no longer in production and it is preferable to use an AI HAT+ or AI HAT+ 2 model instead.

Raspberry Pi AI HAT+

This is available in two models, with either an on-board Hailo-8L NPU or an on-board Hailo-8 NPU.

Raspberry Pi AI HAT+ 2

The most recent, and most powerful option at the time of writing. It features an on-board Hailo-10H NPU.

AI HAT+ 2 also enables you to run generative large language models (LLMs) as seen in "Run an LLM on AI HAT+ 2" on page 101.

INSTALL THE AI HAT+

For instructions on connecting an AI HAT+ or AI HAT+ 2 to a Raspberry Pi 5, see **rpimag.co/aihatinstall**

Update Raspberry Pi OS

Ensure that the Raspberry Pi 5 is running Raspberry Pi OS Trixie with the latest software installed, and that it has the latest Raspberry Pi firmware:

```
sudo apt update && sudo apt full-upgrade -y
sudo rpi-eeprom-update -a
sudo reboot
```

AI KIT AND PCIe

If you're using an AI Kit, we highly recommend that you enable PCIe Gen 3.0. You can skip this for AI HAT+ and AI HAT+ 2 because the setting is automatically applied: **rpimag.co/enablepci3**

Install required dependencies

After updating your Raspberry Pi with the latest Raspberry Pi software and firmware, the following dependencies are required to use the NPU:

▸ The Hailo kernel device driver and firmware.

▸ Hailo RT middleware software.

▸ Hailo Tappas core post-processing libraries.

How you install these dependencies depends on the AI hardware you're using. Choose the appropriate installation option for your AI hardware. The AI Kit and AI HAT+ require a different package (**hailo-all**) than AI HAT+ 2 (**hailo-h10-all**). These packages can't co-exist on the same operating system.

Before you install the software specific to your AI HAT+ hardware, install the required **dkms** package:

```
sudo apt install dkms
```

> **DYNAMIC KERNEL MODULE SUPPORT**
>
> The AI HAT+ hardware requires access to the PCIe port. Both **hailo-all** and **hailo-h10-all** install **hailo_pci**, which is compiled against your currently running kernel during the installation. The **dkms** framework is a standard way to dynamically build kernel modules.

AI Kit or AI HAT+

> To install the required dependencies for the AI Kit or AI HAT+, open the Raspberry Pi Terminal and run **sudo apt install hailo-all**

AI HAT+ 2

> To install the required dependencies for AI HAT+ 2, open the Raspberry Pi Terminal and run **sudo apt install hailo-h10-all**

After installing the required dependencies, you must reboot your Raspberry Pi 5. You can do this from the Raspberry Pi Terminal using the following command:

```
sudo reboot
```

When your Raspberry Pi 5 has finished booting back up again, run the following command to check that everything is running correctly:

```
hailortcli fw-control identify
```

If you see output similar to the following, you've successfully installed the NPU and its software dependencies:

```
Executing on device: 0000:01:00.0
Identifying board
Control Protocol Version: 2
Firmware Version: 4.17.0 (release,app,extended context
                          switch buffer)
Logger Version: 0
Board Name: Hailo-8
Device Architecture: HAILO8L
Serial Number: HLDDLBB234500054
Part Number: HM21LB1C2LAE
Product Name: HAILO-8L AI ACC M.2 B+M KEY MODULE EXT TMP
```

Check the files

The Hailo software installation process will add a number of post processor *json* files. You can view these in the **/usr/share/rpi-camera-assets** (along with any files added with Raspberry Pi Camera or AI Camera installation.

The AI HAT+ files appear with a **hailo_** prefix (as compared the AI Camera files which appear with an **imx500_** prefix). Run the following command:

```
ls /usr/share/rpi-camera-assets/
```

You should see output similar to the following:

```
/usr/share/rpi-camera-assets/hailo_classifier.json
/usr/share/rpi-camera-assets/hailo_pose_inf_fl.json
/usr/share/rpi-camera-assets/hailo_scrfd.json
/usr/share/rpi-camera-assets/hailo_yolov5_personface.json
/usr/share/rpi-camera-assets/hailo_yolov5_segmentation.json
```

```
/usr/share/rpi-camera-assets/hailo_yolov6_inference.json
/usr/share/rpi-camera-assets/hailo_yolov8_inference.json
/usr/share/rpi-camera-assets/hailo_yolov8_pose.json
/usr/share/rpi-camera-assets/hailo_yolox_inference.json
```

These files are passed into the *rpicam-apps* software to perform inferencing with the AI HAT+ hardware. As with AI Camera files (prefixed with **imx500**) you can view the contents of each file:

```
less /usr/share/rpi-camera-assets/hailo_yolov8_inference.json
```

Inside the **.json** file you will find adjustable attributes such as `max_detections`, `threshold`, and `visible_frames`. These can be fine-tuned to match your requirements. As with the AI Camera *json* files, you can copy these files somewhere into your home directory and edit the copies. When you need to use them, change the command you're running so the *json* path component refers to the file in your home directory.

You will also find references to the underlying **.hef** files. Typically, three: **hef_file_10** for AI HAT+2; **hef_file_8** for AI HAT+ with 26 TOPS; and **hef_file_8L** for AI HAT+ with 13 TOPS.

```
    "hef_file_10": "/usr/share/hailo-models/yolov8m_h10.hef",
    "hef_file_8L": "/usr/share/hailo-models/yolov8s_h8l.hef",
    "hef_file_8": "/usr/share/hailo-models/yolov8s_h8.hef",
    "max_detections": 20,
    "threshold": 0.4,
```

These **.hef** files are the *Hailo Executable Format*. They are binary files that package together the weights and computation graph mapped to the Hailo hardware. You can view some details about them with the hailortcli parse-hef command, such as:

```
hailortcli parse-hef /usr/share/hailo-models/yolov8s_h8.hef
```

Here is the output from that command:

```
Architecture HEF was compiled for: HAILO8
Network group name: yolov8s, Single Context
    Network name: yolov8s/yolov8s
```

```
VStream infos:
    Input  yolov8s/input_layer1 UINT8, NHWC(640x640x3)
    Output yolov8s/yolov8_nms_postprocess FLOAT32,
        HAILO NMS BY CLASS(number of classes: 80,
        maximum bounding boxes per class: 100, maximum
        frame size: 160320)
    Operation:
        Op YOLOV8
        Name: YOLOV8-Post-Process
        Score threshold: 0.200
        IoU threshold: 0.70
        Classes: 80
        Max bboxes per class: 100
        Image height: 640
        Image width: 640
```

> **MODEL COMPATIBILITY**
>
> The **.hef** files referenced in the **.json** files end with h8l, h8, or h10. These specify compatibility with AI HAT+ hardware
>
> - **h8**: Compatible with AI HAT+ with 26 TOPS
> - **h8l**: Compatible with AI HAT+ with 13 TOPS or 26 TOPS
> - **h10**: Compatible with AI HAT+ 2
>
> Not all of the post processor files inside **rpi-camera-assets** will work with your variation of AI HAT+. If a file is not working check the reference to the **.hef** file in the **.json** file.

Object detection with AI HAT+ and AI HAT+ 2

After you've verified that everything is correctly installed, you can run camera AI demos using the *rpicam-apps* camera software. This software implements AI demos using a post-processing framework; this software uses pre-trained neural networks to run AI inference on camera frames using the NPU.

To highlight some of the capabilities of the NPU, this section outlines some demos that showcase different models and post-processing stages, such as drawing bounding boxes around objects or pose lines around people. Results are displayed either visually on the live preview window (default) or as text in the Raspberry Pi Terminal.

The following demos use `rpicam-hello`, but you can also use other *rpi-cam-apps*, such as `rpicam-vid` for video recordings and `rpicam-still` for still images. These applications might require you to add or modify some command line options to make them compatible.

The following demos perform object detection using different *YOLO* (You Only Look Once) models. Each demo draws bounding boxes around detected objects and supports optional flags to modify the output, such as `-n` to turn off the viewfinder and `-v 2` to display textural output only.

Different demos have different trade-offs in speed and accuracy. Run the following commands to try each of the demos on your Raspberry Pi 5. These post processor files run on all three variants of AI HAT+ hardware. All of the **.json** files are located in **/usr/share/rpi-camera-assets**. If you'd like to customise them, you can make copies and edit them as described in "Post processor file" on page 171.

Object detection (v6)

The **hailo_yolov6_inference.json** post processor file loads a model that can detect a variety of objects:

```
rpicam-hello -t 0 --post-process-file \
  /usr/share/rpi-camera-assets/hailo_yolov6_inference.json
```

Object detection (v8)

The **hailo_yolov8_inference.json** post processor file uses a model that can also detect objects. It uses a model that has more than double the number of parameters of the v6 version:

```
rpicam-hello -t 0 --post-process-file \
  /usr/share/rpi-camera-assets/hailo_yolov8_inference.json
```

Figure 8-7 shows object detection with the AI HAT+. Compare this against **Figure 8-3** to see how more items you can detect with an AI HAT+ than with an AI Camera.

Figure 8-7   Object detection on AI HAT+ hardware

Pose estimation

This performs 17-point human pose estimation, drawing lines connecting the detected points. This could be useful for analysing gait, or even making a video game that players control by moving their body:

```
rpicam-hello -t 0 --post-process-file \
  /usr/share/rpi-camera-assets/hailo_yolov8_pose.json
```

Segmentation

This model is similar to the object detection examples, but it *segments* objects by drawing a colour mask over each detected object:

```
rpicam-hello -t 0 --framerate 20 --post-process-file \
  /usr/share/rpi-camera-assets/hailo_yolov5_segmentation.json
```

Figure 8-8 shows how this model draws translucent overlays over objects.

Figure 8-8 Image segmentation partitions the image into sections to detect different objects and boundaries

Hailo Model Zoo and AI Software Suite

You can expand the functionality of AI HAT+ by downloading models from the Hailo Model Zoo, which forms part of the Hailo AI Software Suite (**rpimag.co/hailosuite2026**). This software is available from the Hailo Developer Zone (see **rpimag.co/hailosoftware**).

The AI Software Suite consists of:

- ▸ Dataflow Compiler (model conversion and compilation to Hailo binary format)

- ▸ HailoRT (runtime environment and driver for running networks and interacting with Hailo devices)

- ▸ Model Zoo (pre-trained vision models to run and evaluate on Hailo devices)

- ▸ Model Zoo GenAI (pre-trained GenAI models, including LLMs and VLMs, to run on Hailo devices)

- ▸ TAPPAS (GStreamer-based pipeline infrastructure for building vision AI applications)

- ▸ Hailo Apps (ready-to-run vision AI application examples built on top of TAPPAS)

The full AI Software Suite requires Ubuntu running on x64-based hardware and recommends 32GB RAM and an Nvidia-class graphics card. We won't get into that in this book, because there is plenty you can do without the full suite.

Chapter 9

Recognise video and image content on the CPU

Use YOLO to recognise and classify objects
without an AI accelerator

YOLO (You Only Look Once) is a powerful object detection model created
by Ultralytics that enables you to identify content in images and video
from the command line and Python. From here you perform classification
and respond to images, videos with your code.

YOLO forms a powerful means of identifying objects that the Raspberry
Pi board can react to. You can use it with sensors and actuators connected
to Raspberry Pi to real-time identification and reaction. Many of the post-
processor models used by AI Camera and AI HAT+ hardware are based on
YOLO technology.

YOLO26 is the latest version available at the time of writing, and is custom
built for speed, accuracy, and versatility. You can use YOLO out of the box
or train your own dataset on it.

In this chapter, we'll run the Ultralytics YOLO framework on a Raspberry
Pi 5 with image and video files. You don't need a Raspberry Pi Camera for
this, but a reasonably powerful Raspberry Pi will help.

Start with Docker

Set up Docker on your Raspberry Pi (see "Set up Docker" on page 24 for instructions). Now that we have Docker installed we can pull the pre-built Ultralytics YOLO26 image for Raspberry Pi.

Execute the following command to pull the Docker container and run on Raspberry Pi. At the time of this writing, the container is based on an **arm64v8/ubuntu** docker image which contains Ubuntu 24.04 with a Python3 environment.

```
docker pull ultralytics/ultralytics:latest-arm64
docker run -it --ipc=host ultralytics/ultralytics:latest-arm64
```

YOU ARE ROOT!

The docker container runs as root. This is indicated by the username (**root**) and the **#** symbol at the end of the prompt. When you are running as a normal (unprivileged) user, you'll see a **$** symbol at the end of your prompt instead.

Remember that you don't need to preface commands that normally require elevated privileges with **sudo**. But be careful with the commands that you do run.

You will now be running a Docker instance, and your shell prompt will be replaced with the following:

```
root@44214a2e5000:/ultralytics#
```

Enter **ls** to take a look at the files inside the docker instance. Note the presence of **yolo26n.pt** in the directory. We will be using this PyTorch file in the next step.

Use YOLO26n from the command line

To perform image detection on Raspberry Pi we convert the YOLO26n model from PyTorch format to NCNN (**rpimag.co/ncnn**).

Out of all the model export formats supported by Ultralytics, NCNN delivers great inference performance when working with Raspberry Pi devices because NCNN is highly optimised for mobile and embedded platforms (such as ARM architecture).

Convert from PyTorch to NCNN

The YOLO26n model is in PyTorch format, so we'll convert it to NCNN so we can run inferencing with the model. Let's convert the file from the command line:

```
yolo export model=yolo26n.pt format=ncnn
```

This takes **yolo26n.pt** file in our directory and creates a **yolo26n_ncnn_model** directory. Take a look inside with:

```
ls yolo26n_ncnn_model
```

You will see several files that make up the NCNN model:

```
metadata.yaml   model.ncnn.bin   model.ncnn.param   model_ncnn.py
```

Use the NCNN model to perform object detection

We now use the **yolo predict** command with two options: **model**, which is pointed to the **yolo26n_ncnn_model** folder and **source**, which in this instance is a **bus.jpg** image on the Ultralytics website:

```
yolo predict model='yolo26n_ncnn_model' \
  source='https://ultralytics.com/images/bus.jpg'
```

This code pulls an image of a bus from the Ultralytics website and per-forms object inference on it. Look for these lines in the output:

```
image 1/1 /ultralytics/bus.jpg:640x640 4 persons,1 bus, 194.4ms
Results saved to /ultralytics/runs/detect/predict
```

This informs us that there is one image, it's called **bus.jpg**, is 640✕640 res-olution, and inference saw 4 people and 1 bus. The model took 194.4ms to run. It also lets us know where the prediction was saved.

Use **ls** to look in the **/ultralytics** folder again. You will see the **bus.jpg** file that was downloaded. Enter:

```
ls runs/detect/predict
```

And you will see another **bus.jpg** file. This one has bounding boxes around it for viewing.

There's no easy way to view this file inside the Docker instance as we don't have any graphics programs or a windowing system. So, this is a good chance for us to learn how to add and remove files from our Docker instance to the desktop.

Copy files to and from Docker

Don't exit the Docker session, because Docker might decide to shut down the running instance, in which case the image and model we just used will be lost. Instead, open a new, second, terminal instance to reach inside Docker from our user account. First, we need to know what the name of our docker instance is:

```
docker ps
```

This should give us information on our running docker instances (this is a shortcut for **docker container ls**). By default, the command returns some output columns that we don't need. You can control what it displays with the **--format** argument:

```
docker ps --format "table {{.Image}}\t{{.Names}}"
```

It will return something like this:

```
IMAGE                                   NAMES
ultralytics/ultralytics:latest-arm64    keen_bohr
```

The name of our container is created randomly from two words and is under NAMES. In our case it is **keen_bohr** but yours will be different. We use this name to reach inside the container and grab the **bus.jpg** file using Docker's built in copy (**cp**) command with source and destination paths:

```
docker cp keen_bohr:/ultralytics/bus.jpg ~/Downloads
```

This copies the bus.jpg from our default **ultralytics** folder (where the container starts) to our **Downloads** folder. You can open it with the following command (see **Figure 9-1**):

```
xdg-open ~/Downloads/bus.jpg
```

This is the test file downloaded from Ultralytics. More interesting is the processed file after we performed inference on it. These files live inside the **runs/detect** folder inside the home directory within the container.

```
docker cp keen_bohr:/ultralytics/runs/detect/predict/bus.jpg \
    ~/Downloads/bus_predict.jpg
```

> ### MULTIPLE PREDICTION FOLDERS?
>
> If you have run more than one prediction it may be saved in another folder such as **predict-2** or **predict-3**. Take a look using `ls runs/detect` inside the Docker container to find your **bus.jpg** file.

Open this file by running `xdg-open ~/Downloads/bus_predict.jpg` at the command line or using the File Manager. You'll see the bus and four people with bounding boxes around them, as shown in **Figure 9-2**.

Figure 9-1 A test image of a bus used by Ultralytics to demonstrate YOLO's object detection ability

Use your own files

You can replace the source reference in our **yolo predict** command with the URL for a different online image (note that some services will block it). More importantly you can use **docker cp** to send image files to the docker instance for inferencing.

We've got an image of our cat sitting on a bed (see **Figure 9-3**). We can send it to our Ultralytics instance and perform inference on it. In our terminal (outside of the Docker instance) enter:

Figure 9-2 Ultralytics' standard demonstration detects a bus and four people in the image

```
docker cp cat.jpg keen_bohr:/ultralytics/cat.jpg
```

Replace **keen_bohr** with the name of your Docker instance.

Head back to your terminal window running the Docker instance, and perform inference on the image you just sent over:

```
yolo predict model=yolo26n_ncnn_model source='./cat.jpg'
```

Copy the resulting file from Docker and open it, as shown in the previous section, and you should see something like **Figure 9-4**. The file may be in the **/ultralytics/runs/detect/predict-2** folder or similar. It has detected a cat (with 88% certainty), a cup (31%) and a bed (52%).

Figure 9-3   A photograph of a cat sitting on a bed

Video files

We can also perform inference on video files, both online and uploaded directly to our Docker instance with the **yolo26n-seg.pt** model. Use this command to perform object detection on a YouTube video. First, we need to get the model from Ultralytics' GitHub site. Run this command from inside the Docker image:

```
wget https://rpimag.co/yolo26n-seg.pt
```

That will download a version of the **yolo26n-seg.pt** model that works with the examples in this chapter. As of this writing, that's **v8.4.0**.

As with the image model, you must convert it to the NCNN format for improved performance on Raspberry Pi:

```
yolo export model=yolo26n-seg.pt format=ncnn
```

Figure 9-4 An image of a cat that YOLO26 has performed image inferencing on and added a bounding box and labels

MORE MODELS!

Discover more models to download and experiment with on the Ultralytics GitHub site: **rpimag.co/ultralyticsgit**.

Let's run the model on a video from the Raspberry Pi YouTube channel:

```
yolo predict model=yolo26n-seg_ncnn_model imgsz=320 \
    source='https://www.youtube.com/watch?v=oo5wb4LDWW4'
```

As it runs you will see the frame detections ouput to the command line:

```
0: 640x640 1 person, 224.4ms
0: 640x640 2 persons, 201.0ms
...
0: 640x640 3 persons, 1 bottle, 3 chairs, 217.7ms
0: 640x640 3 persons, 1 suitcase, 1 bottle, 3 chairs, 211.0ms
```

When the inference has finished you can download the video with the exported frames and bounding boxes.

The video will be saved in your Docker container under **runs/segment/predict**. Head to another terminal window (outside of the Docker instance) and run these commands to first rename the file, and then to copy it to your Raspberry Pi OS filesystem:

```
docker exec keen_bohr \
  mv runs/segment/predict/https___www.youtube.avi \
     runs/segment/predict/output.avi

docker cp \
  'keen_bohr:/ultralytics/runs/segment/predict/output.avi' \
  ~/Downloads/output.avi
```

Replace **keen_bohr** with your own container (enter **docker ps** to get the name). You can then view the video with VLC. **Figure 9-5** shows one of the frames from this video.

FOLDER NAMING

The predict folder may be named **predict-2**, **predict-3**, and so on if you have run multiple inferences. Check on the exact file path inside the docker instance with `ls runs/segment/predict`.

Python Usage

Performing object detection from the command line is fun, but using inference models in Python is where you can start to explore the power of AI in coding. At this point you can start to integrate the vision detection models into your own code and connect it to sensors and actuators.

Let's look at how to download the **bus.jpg** file from Ultralytics, perform inference, and output the results to the terminal window, all in Python. You may want to install a text editor, such as nano or vim, to work on your Python code. You can do this with **apt** inside the Docker instance.

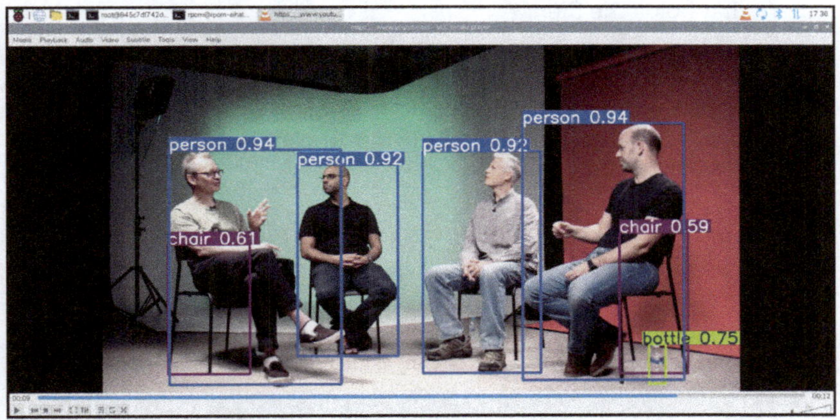

Figure 9-5   An image from a YouTube video with image detection performed on it

```
apt update
apt install vim -y # or nano
```

You could also clone the GitHub repository for this book (see *Welcome* on page v) in your Docker image. You can find the program in the **ch09** directory as **bus.py**:

```python
from ultralytics import YOLO

# Load a pretrained YOLO model
model = YOLO("yolo26n_ncnn_model", task="detect")

# Perform object detection on an image
results = model("https://ultralytics.com/images/bus.jpg")

# Visualize the results
for result in results:
    result.show()
```

If you checked out the GitHub repository, you should change to the directory that holds the file:

```
cd ai-projects-raspberrypi/ch09
```

And save it to the ultralytics folder in your Docker instance. Now run the code:

```
python bus.py
```

This program will return the following:

```
Loading /ultralytics/yolo26n_ncnn_model for NCNN inference...

Found https://ultralytics.com/images/bus.jpg locally
  at bus.jpg
image 1/1 /ultralytics/ai-projects-raspberrypi/ch09/bus.jpg:
  640x640 4 persons, 1 bus, 102.6ms
Speed: 8.7ms preprocess, 102.6ms inference, 2.8ms postprocess
  per image at shape (1, 3, 640, 640)
```

Further reading

Now that you have Ultralytics YOLO up and running you can start working through the Python documentation to integrate it into your projects. Bookmark **docs.ultralytics.com** and take a look at the following Ultralytics resources to continue your journey:

Ultralytics

rpimag.co/yolorpi

NCNN

rpimag.co/yoloncnn

Python Usage

rpimag.co/yolopython

Chapter 10

Train a Rock, Paper, Scissors AI model

Learn how to train your own models that can run on the Raspberry Pi AI Camera

At the heart of each AI system is its model. This is the engine that makes the AI work. There are different types of models for different tasks. Convolutional Neural Networks (CNNs) are great for images; Large Language Models (LLMs) are great for handling text, and Reinforcement Learning (RL) models are great for environmental interaction, such as teaching a robot to walk.

In this tutorial we are going to retrain MobileNet V2. This is a CNN designed for image classification. The original model is designed to detect thousands of different objects. We are going to retrain it to detect just three: hand shapes forming rock, paper, or scissors gestures.

Physical AI

Our trained model is practical and can be used in the physical world. You can use this code to create custom detection models for all kinds of industrial and real-world purposes. You could train it to detect faulty/working items on a production line, or stock levels on shelves; weeds vs crops in a field. Or just about anything where you need a camera to look at something and answer 'is it this, or this, or this?'

This chapter helps us understand (and experience) the two key stages of an AI model:

Training

> This is the computational heavy lifting where you adjust the parameters in a model until it can reliably map inputs (images of hands, in this instance) to outputs (labels of 'rock', 'paper', or 'scissors' based on the shape of the fingers).

Inference

> This is where you deploy the model to make predictions or decisions. Computationally it's much lighter than training and inference is easily run locally on smaller devices, such as Raspberry Pi.

Despite us training on a Raspberry Pi we are leaning on the shoulders of giants, in this case the mighty MobileNet V2 architecture (**rpimag.co/huggingface-mobilenet**). Rather than train all the thousands of layers at once, we slice off the final categorisation layer and train this layer. Thanks to this clever approach, it is perfectly possible to train a model like this on a Raspberry Pi (it helps to have a 16GB model). It's also more functional because we maintain the body of the trained model (trained to detect shapes, curves, colours, and so on with millions of images).

It is still a bit time consuming. Expect to spend around 30 minutes waiting for the model to train on a Raspberry Pi 5 with 16GB of RAM. It took us less than five minutes on an x86_64 computer with an NVIDIA graphics card. If you're lucky enough to have access to both, we have provided a separate GitHub branch for you to test each out. There is also a version on Google Colab that provides access to Tesla T4 graphics cards.

WHAT YOU'LL NEED

- Raspberry Pi 5
- Raspberry Pi AI Camera
- Raspberry Pi OS
- Docker

Hardware requirements

For this tutorial you will need a Raspberry Pi computer to train the model and an AI Camera to run the converted package. The AI Camera requires any Raspberry Pi with a CSI (Camera Serial Interface) connector. Because the inference is being handled at a hardware level on AI Camera, it works with good results even on a Raspberry Pi 3B or Raspberry Pi Zero 2 W. Using later models allows higher resolution streaming at higher frame rates.

Our program has complex software requirements and there are two ways to run it:

▸ Via Python code running inside a Docker container on Raspberry Pi OS. This will take longer but runs locally.

▸ Inside a Jupyter Notebook environment (**jupyter.org**) running on Google Colab (**colab.research.google.com**, shown in **Figure 10-1**). This provides cloud access to T4 GPU hardware and performs training in a matter of minutes.

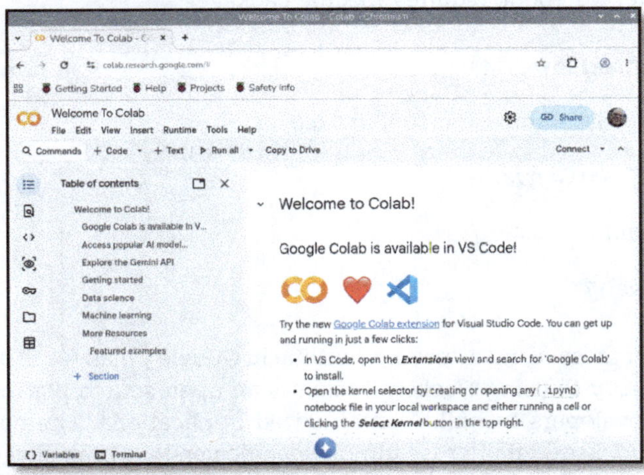

Figure 10-1   Google Colab enables you to access powerful compute hardware from your Raspberry Pi

Jupyter Notebook

Project Jupyter is a web-based interactive environment for data science and machine learning. It enables developers to combine Python code with Markdown cells. This lets you combine code with detailed documentation, which explains the code *in situ* and makes for a powerful educational environment.

Code in Jupyter Notebook is split into cells. These can contain Python code, shell commands, images, audio, video and even interactive widgets and plots. It's a powerful way to combine code with descriptive material that makes it tremendously useful for explaining how code works.

At its heart, though, is Python code and you will train our model with just Python and the appropriate frameworks.

Package requirements

Our Jupyter Notebook requires specific packages:

- TensorFlow 2.14.0

- Model Compression Toolkit 2.2.0

- IMX500 Converter

- NumPy 1.26.4

- Java 17

We can run these using a specific runtime in Google Colab (2025.07) or on our Raspberry Pi using Docker. Docker is an open source platform that enables developers to build containerized applications. The code that runs in a Docker container is built on specific versions of packages, running on specific operating systems.

These are all installed in a Dockerfile allowing us to concentrate on running the code. You can use Docker to train your model locally on Raspberry Pi or a computer with GPU support if you'd like.

It is faster to use Google Colab to train the model and deploy your model. This is also a great educational environment that is useful to learn, but you are sending your training data to the cloud, which has implications for privacy, especially if you plan to move beyond our training dataset and start using your own images.

If you'd rather set up your training model to work locally, you'll need to install Docker first. See "Set up Docker" on page 24 for information on setting up Docker on your Raspberry Pi.

Option 1. Set up Google Colab

Sony has provided a version of this code in Google Colab. This virtual environment enables you to bypass Docker and run the Jupyter Notebook in a web interface with GPU support.

Open the Chromium web browser in Raspberry Pi OS (or Chrome on another computer) and visit Google Colab: **rpimag.co/mobilenetcolab**. You will need to sign in with your Google Account. The training notebook will appear as shown in **Figure 10-2**.

Before we start with the notebook we need to adjust the runtime. There is a conflict between the TensorFlow 2.14.0 requirement of our Notebook and the latest versions of Google Colab's runtime (which supports TensorFlow 2.16.0 or later). To fix this we first swap our runtime version to 2025.07.

Click the **Runtime** menu and select **Change runtime type**. Change the **Runtime Version** setting to 2025.07 and click **Save**. **Figure 10-3** shows how to change the runtime.

Now click **Connect** in the top-right of the interface. It may take a few minutes to connect to the runtime because we are running an older version. Wait until it finishes connecting, at which time the **Connect** button will be replaced with little graphs showing RAM and disk usage.

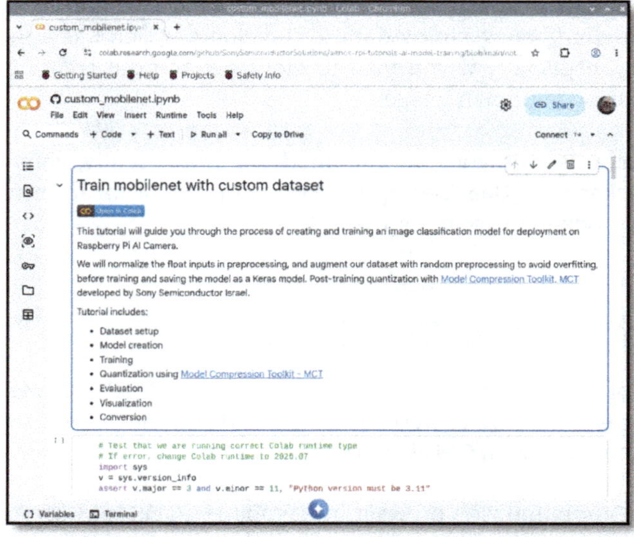

Figure 10-2   The Rock, Paper, Scissors training notebook in Google Colab

Run the code in Google Colab

If you're eager to get the code running, this section will provide a quick start. We'll go into the notebook in more detail in "Exploring the notebook code" on page 214. Run each cell individually by highlighting it and pressing **SHIFT+ENTER**. This is a good way to go through each code block one at a time.

COLAB WARNING

The first time you run a code cell Colab will display a warning that the notebook was not authored by Google. Click **Run anyway** because we know the provenance of this notebook. However, you should always review the source code before running other notebooks.

Next to each cell is a marker displayed with square brackets: **[ ]**. As the code runs it will show a number inside, such as **[1]**, with an animated

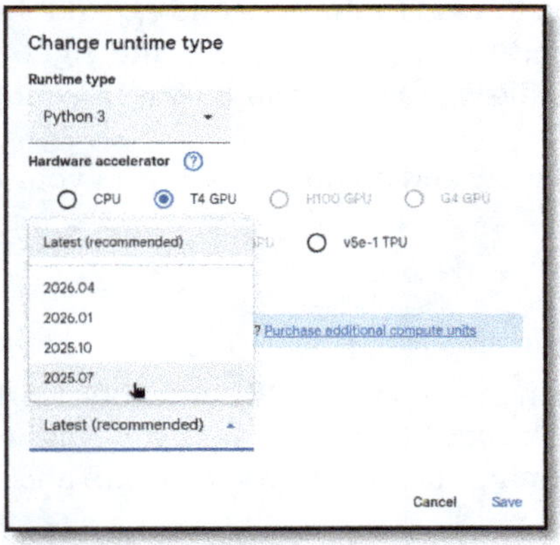

Figure 10-3 Switch to an older runtime to access the correct version of TensorFlow

spinner just to the right of the number. This is the code execution prompt and will let you know which code is running, and what ran in which order.

You can run all the code by clicking **Run All** in the menu bar or selecting the **Runtime** menu and choosing **Run All Cells** but don't do that right now because we'll need to restart the session shortly.

After a code block has run, the spinner will disappear, and a green tick will appear next to the prompt along with the number of seconds it took to finish running.

MANDATORY RESTART

After running the TensorFlow code cell you will be asked to restart the session. Click the **Restart Session** button. Then click the **Runtime** menu and select **Run Cell and Below** to continue.

It should take between 10 and 30 minutes to train the model in Google Colab. When the code has finished running you will get a **packerOut.zip** file that we will convert into a **network.rpk** file (this will run on our AI Camera to detect hand gestures).

When all the code cells have finished running, we can get our **packer-Out.zip** file. Click on **Files** in the sidebar and expand the converted folder (see **Figure 10-4**). Locate the **packerOut.zip** file, right-click on it and select **Download**. The file will be saved into your **Downloads** folder.

Open a terminal window and confirm that it is in your **Downloads** folder (see **Figure 10-5**):

```
ls  ~/Downloads
```

If you'd like to learn more about how the notebook works, check out "Exploring the notebook code" on page 214). If you want to move right onto the next step of running the model, you can skip ahead to "Package the file" on page 232.

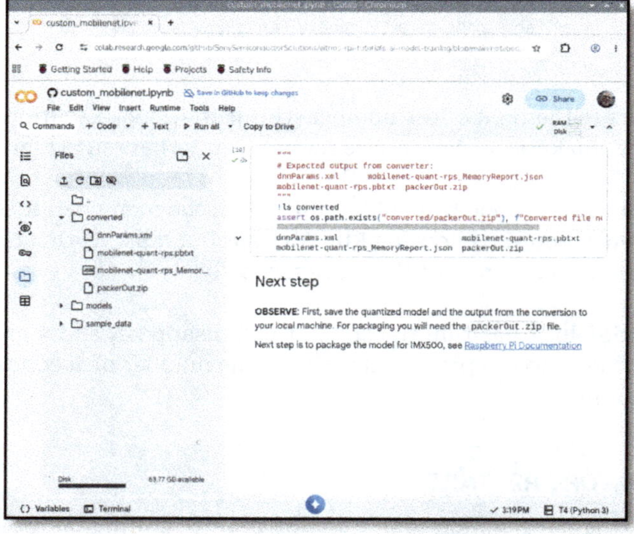

Figure 10-4   The packerOut.zip file in our Files directory

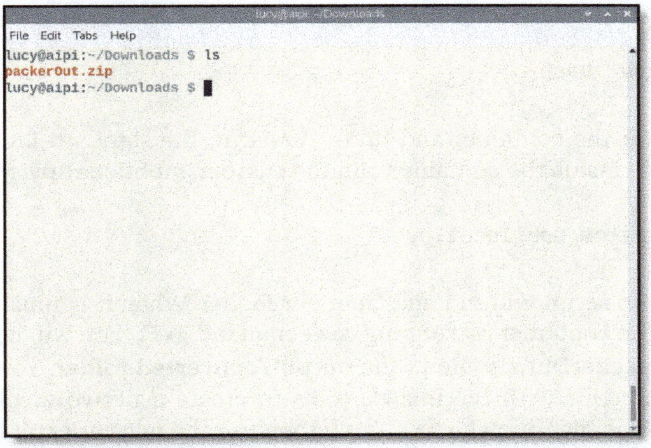

Figure 10-5   Confirming the download completed successfully

Option 2. Run in Docker

If you want to run this code on Raspberry Pi, you can use a Python file we have adapted on a local Raspberry Pi installation. First, we need to clone the code from GitHub (see *Example code* on page ix). When you're done, navigate into the **custom_mobilenet** directory:

```
cd ai-projects-raspberrypi/ch10/custom_mobilenet
```

Ensure you have Docker set up (see "Set up Docker" on page 24). Once Docker is up and running, you can build the Docker Image:

```
docker build -t mobilenet .
```

Now run the image as a Docker container with the **-it** flags for interactive mode with a terminal. We give the container the name **mobilenet** so we can access it again and we also attach two volumes: **models** and **converted**. This ensures we can access the models once they have been trained.

Run the following command:

```
docker run -it --name mobilenet \
  -v "$(pwd)/output/models:/app/models" \
  -v "$(pwd)/output/converted:/app/converted" \
  mobilenet bash
```

You'll enter the container and find yourself in the standard Linux Bash shell. Once inside the container, run the **custom_mobilenet.py** script:

```
python custom_mobilenet.py
```

The Python script will run and train our model. When it is finished, you can exit the container by running the command **exit**. You will be able to find the **packerOut.zip** file in the **output/converted folder**. You'll learn how to use this with the imx500 tools to create a **network.rpk** file in "Package the file" on page 232. You'll then use the **network.rpk** file with the Raspberry Pi AI Camera to detect hands.

Before we get into how to deploy and run the model, let's take a look at what the code is doing to *create* the model.

Exploring the notebook code

The code is split into several stages. One of the handy things about Jupyter Notebooks is that code and text is placed into cells, known as Code blocks and Text blocks. Navigate these cells using the **UP** and **DOWN** arrow keys and use **SHIFT+ENTER** to run each code cell.

Although the **custom_mobilenet.py** file contains essentially the same code as the notebook, you may find it easier to browse and study the notebook in Google Colab (**rpimag.co/mobilenetcolab**) even if you don't plan to run it there.

EDIT MODE

If you press **ENTER** in a formatted informational cell, you will switch to editing mode and see the HTML for the cell. Press **ESC** or click **Close** to exit this mode.

The code contains the following cells:

- **Train mobilenet with custom dataset** — An information box that explains the notebook followed by a Python code cell that checks we are running the correct Colab runtime type.

- **Installation** — Installs TensorFlow 2.14.0 and Java. After this step you will be asked to restart the kernel. This is to sync Jupyter kernel with the virtual machine.

- **Settings** — Initial setup determining constants and file locations.

- **Dataset** — We will use the RPS (rock, paper, scissors) dataset from TensorFlow datasets. This comes in two parts: a training set and testing set.

- **Pre-processing and augmentation** — Images are sent through in batches, here we set the range, adjust the image sizes, and slightly adjust each image to improve the performance.

- **Keras Mobilenetv2 model for transfer learning** — We bring in the MobileNetV2 model and freeze its base layers. This saves us from training an entire model and enables us to concentrate on the head (the final classification layer).

- **Train** — This is the computationally intensive process of training a model on the dataset. We pass through the data up to 20 times, with each pass known as an 'epoch'.

- **Quantization** — This is a process of optimization that maps the larger values into a smaller set to reduce the complexity of it.

- **Evaluation** — Here we check the accuracy of the model.

- **Visualize detections** — This section of code displays a selection of test images and inference results.

- **Conversion** — Output the quantized Keras model to a **zip** file. We can then convert this into a **network.rpk** file for use on AI Camera.

- **Next step** — The final cell gives us information on saving the **packerOut.zip** file and how to package it for the IMX500 sensor inside Raspberry Pi AI Camera.

The code is somewhat lengthy, and like most things in AI-world, quite gnarly on first pass. But it is tremendously valuable to learn how this process works, both for the educational value and for the ability to train models on custom datasets with real-world implications.

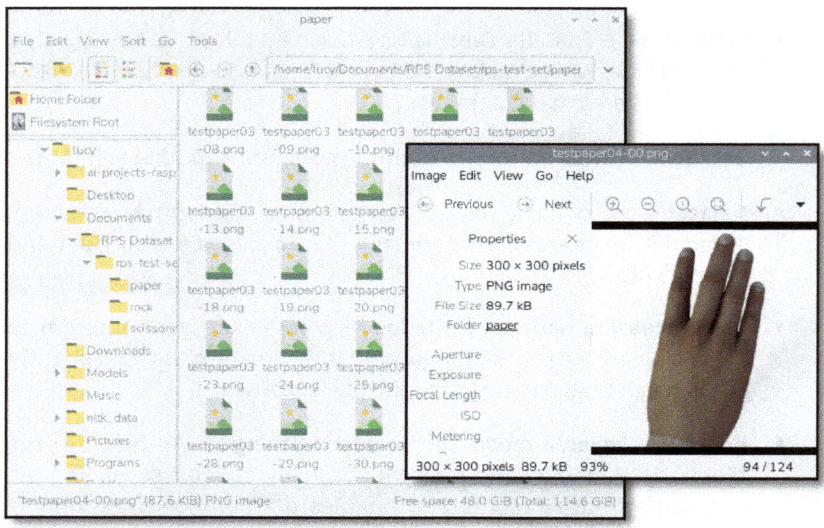

Figure 10-6   The folder containing test images of hands showing a paper gesture

RPS DATASET

You can view the **rock_paper_scissors** dataset information on **tensorflow.org**. Downloading the raw images gives you an understanding of the images used and the three categories: train, test, and validation: **rpimag.co/rpsdataset**.

Figure 10-6 shows the images in a folder, and **Figure 10-7** shows one of the training images (with the hand in a "paper" gesture).

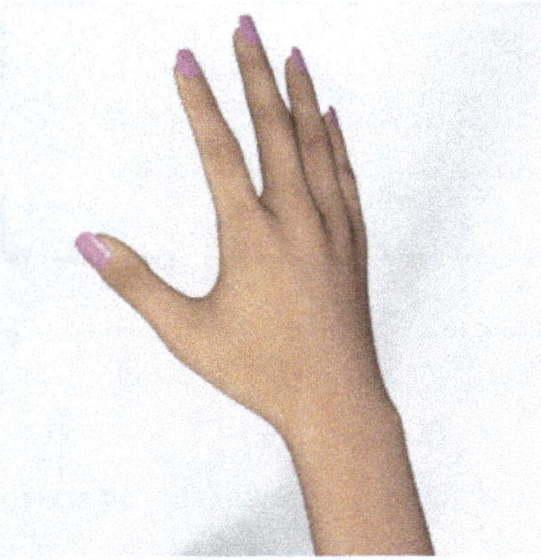

Figure 10-7
One of the thousands of training images

Installation

This code block uses **!pip** to install TensorFlow, Model Compression Toolkit, and IMX500 Converter.

```
!pip install tensorflow~=2.14.0 \
    model-compression-toolkit~=2.2.0 \
    imx500-converter[tf]
```

The **!** is a special Jupyter/Colab syntax that lets you run shell commands directly from a notebook cell. Without it, Jupyter Notebook, would consider **pip** to be a Python variable.

After installing TensorFlow you will be asked to restart the session. Click the **Restart Session** button as shown in **Figure 10-8**.

This cell also throws up a few red errors, but we can ignore them and move on to the cell that starts with "**# Converter requires java**". This cell installs Java 17 instead of our runtime's default version of Java 11.

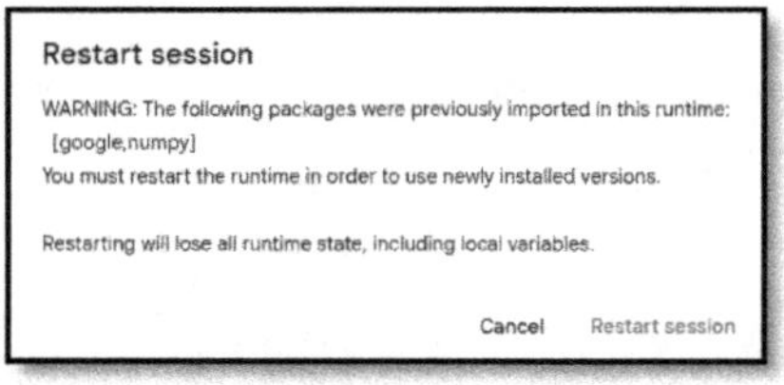

Figure 10-8 Restart the session after installing TensorFlow

When it has finished installing Java you should see the following:

```
Found Java version: 11.0.27
Java 17 is not installed. Installing correct version...
Java installed
```

Once our runtime is correct and all the software is installed, we can move on to the Python code.

Settings

This next code block sets up our `MODEL` and `MODEL_KERAS` variables that we will use later in the program. It creates a directory called **models** in our runtime to store the produced image model files. It also creates two constants to store the location of our trained model in **.keras** format.

> **KERAS**
>
> Keras is an API (application programming interface) that sits on top of TensorFlow. Keras is what we use to interact with the images and train the inference model, while TensorFlow handles the algebraical numerical crunching under the hood: **keras.io/api**.

We also create two constants related to our model. Because we have thousands of images in our training set, `BATCH_SIZE` determines how many we send through the system at once: 32. **IMAGE_SHAPE** is set to (224, 224) which matches the dimensions of our trained model.

Dataset

We have two cells in this section. First, we have some **import** statements that bring in **tensorflow**, **numpy**, and **keras** packages.

The second code block lines up the Rock, Paper, Scissors dataset for download. We start off with our **dataset_name** var, which we set to **rock_paper_scissors**. The **(train_ds, validation_ds), info** variables hold the information returned from the **tfds.load** module.

Our **print()** statement outputs the contents of **info**, and it's worth looking at because this outlines our dataset. This will return:

```
tfds.core.DatasetInfo(
    name='rock_paper_scissors',
    full_name='rock_paper_scissors/3.0.0',
    description="""
    Images of hands playing rock, paper, scissor game.
    """,
    homepage=\
     'http://laurencemoroney.com/rock-paper-scissors-dataset',
    data_dir=\
     '/root/tensorflow_datasets/rock_paper_scissors/3.0.0',
    file_format=tfrecord,
    download_size=219.53 MiB,
    dataset_size=219.23 MiB,
    features=FeaturesDict({
        'image': Image(shape=(300, 300, 3), dtype=uint8),
        'label': ClassLabel(shape=(), dtype=int64,
                            num_classes=3),
    }),
    supervised_keys=('image', 'label'),
    disable_shuffling=False,
    nondeterministic_order=False,
    splits={
```

```
        'test': <SplitInfo num_examples=372, num_shards=1>,
        'train': <SplitInfo num_examples=2520, num_shards=2>,
    },
    citation="""@ONLINE {rps,
    author = "Laurence Moroney",
    title = "Rock, Paper, Scissors Dataset",
    month = "feb",
    year = "2019",
    url = \
      "http://laurencemoroney.com/rock-paper-scissors-dataset"
    }""",
)
class_names: ['rock', 'paper', 'scissors']
```

This tells us the provenance of the images, the image dimensions: 300, 300, 3 (height, width, channels). These will be resized to 224×224 to match the MobileNet V2 expectation. It also tells us `num_classes=3` (rock, paper, scissors). We also have the split of train/test images (which is 2520 for training, and 372 for testing).

It is important not to use the same data for training as testing to prevent overfitting and ensure that the model operates correctly on fresh data (here, that will be when you place your hand in front of the camera).

We also set two more variables: `class_names` and `num_classes` (set to the `len()` of `class_names`). And we print out `class_names`, which returns a list: `['rock', 'paper', 'scissors']`.

We are now ready to start processing the dataset for our training session.

Preprocessing

Before we train our model, we need to set up a preprocessing pipeline. This preps the conversion process for our data (in this case images) into a friendly format for our neural network.

It helps to think of the `def preprocess_data(image, label):` function as a muscular form of type casting.

This bit of code first casts 8-bit images into 32-bit floating point numbers:

```
image = tf.keras.applications.mobilenet_v2.preprocess_input(
    tf.cast(image, tf.float32))
```

Our dataset (a series of images) is a series of unsigned 8-bit integers under the hood. These are values between 0-255. For example, the RGB of a pixel will contain three 8-bit values, such as:

```
[0, 127, 255] # sky blue pixel, dtype=uint8
```

After **tf.cast** these will be floating point values:

```
[0.0, 127.0, 255.0] # dtype=float32
```

Then a further conversion takes place with the function. MobileNet is trained on images that are normalised in a scale between **[-1, 1]**. So **mobilenet_v2.preprocess_input()** rescales these float32 values to this range. Our sky blue pixel now looks like this:

```
[-1.0, -0.0039, 1.0]  #dtype=float32)
```

Next, we will resize our images as they come through the pipeline to 224×224 pixels.

```
tf.image.resize(image, IMAGE_SHAPE)
```

We set the **IMAGE_SHAPE = (224, 224)** constant in our Settings section earlier. Finally, we return **(image, label)**. Label is unchanged because it contains the value 0, 1, or 2 for rock, paper, or scissors.

Augmentation

Here is a funky bit of code with a cool cyberpunkesque name. Augmentation, for those video game buffs, is where your character gets power-ups to become stronger. And that's exactly what this does to our images. Here we can see two functions:

```python
data_augmentation = tf.keras.Sequential([
  tf.keras.layers.RandomFlip('horizontal'),
  tf.keras.layers.RandomRotation(0.2),
])
```

These settings randomly flip each image left to right with a 50% chance, and randomly rotate each image up to 0.2 radians (around 11 degrees).

Every time we iterate over the dataset, the images will be slightly different due to this randomisation. This effectively inflates our dataset size using the same images, which makes our trained model stronger.

We then have three extra dataset functions:

```python
train_ds=train_ds.map(lambda x, y: (data_augmentation(x), y))
train_ds=train_ds.prefetch(tf.data.experimental.AUTOTUNE)
validation_ds=validation_ds.prefetch(
    tf.data.experimental.AUTOTUNE)
```

The first function maps the `data_augmentation` function to **x** (image) and **y** (label) so we get our new dataset. The next two functions prefetch our training and validation datasets to speed things up. This loads the next augmented dataset while we are training on the current one.

Keras MobileNet V2

MobileNet is a Convolutional Neural Network (CNN) trained on the ImageNet (**image-net.org**) database (nicknamed the 'Big Zoo') which has over a million images in 1,000 categories. The training process has taught it visual patterns: edges, textures, and shapes. This pretrained model has layers that are then reused in our rock, paper, scissors model.

We take the model and remove the final classification layer (the 1,000 categories) and replace it with our three categories (rock, paper, and scissors). We also freeze the pretrained layers so their weights don't change. So, we train a new layer on the new images. Later we fine-tune the model by unfreezing the layers.

By standing on the shoulders of a giant, we can train a model using thousands of images instead of the millions in ImageNet. And training is much faster because we are using a pre-learned network instead of starting from scratch. After a few imports we set our base model:

```
base_model=MobileNetV2(weights='imagenet', include_top=False,
                       input_shape=IMAGE_SHAPE+(3,))
```

This assigns **base_model** to MobileNetV2 using the pre-trained **imagenet** weights and without the classification layer (**include_top=False**). It also sets the **input_shape** to the **IMAGE_SHAPE** constant we set earlier (224×224) and adds on the three channels for RGB.

> ### TOP AND LAST
>
> The top layer is the last layer of the model. The one that originally had 1,000 classifications and we will replace it with three. In a CNN there are two main parts:
>
> - **Feature extractor** — This is the stack of layers that turns pixels into patterns. Early layers will detect rough details like edges and colours; later layers detect curves and shapes.
>
> - **Classifier** — This is the 'top' or 'head' of the model. This layer takes the extracted features and flattens them into a classification score.

Now that we have removed our top layer, we add on our custom three layers:

```
# Add custom classification layers on top
x = base_model.output
x = GlobalAveragePooling2D()(x)
predictions = Dense(num_classes, activation=tf.nn.softmax)(x)
```

This piece of code is passing our **output** tensor from MobileNetV2 into **x**, which is then passed through **GlobalAveragePooling2D**.

This takes the last layer of MobileNetV2 (a 7×7×1280 tensor) and collapses the 1280 channels to one that is averaged by the 7×7 value. So, we just get the one 1280-D vector. Global Average Pooling is a robust optimisation that has better performance than the more traditional Flatten.

Finally, we create our custom head storing in a **predictions** object. We use the **Dense()** class with our **num_classes** variable (which is **3**, remember: rock, paper, scissors) and a **tf.nn.softmax** activation function. This converts the raw numbers (logits) into positive floats between 0 and 1 (such as 0.97, or 0.3). This makes it easier for our classifier to decide which is more probable.

WHY 7×7×1280?

MobileNetV2 is a convolutional neural network. This is a series of layers that the image is passed through, each one increases the number of channels which then need squeezing back down. So, the number of channels increases from 16, 24, 32, and up to 1280. Meanwhile the 224×224 image is divided down to 112×112, 56×56, 28×28, 14×14, and finally 7×7×1280.

If this isn't making much sense, then take a look at 3Blue1Brown's YouTube Video: *But what is a neural network?* **rpimag.co/whatisaneuralnetwork**.

Our next piece of code creates the model using the **base_model.input** and our predictions model:

```python
# Create the full model
float_model = Model(inputs=base_model.input,
                    outputs=predictions)


# Freeze layers in the base model
for layer in base_model.layers:
    layer.trainable = False
```

We freeze the layers of the base model by setting **layer.trainable = False** for all items in **base_model.layers**. This ensures that we only train the final layer, and not the whole model on our rock, paper, scissors images. MobileNet V2 was trained on millions of images and we want to retain its muscular training.

Finally, our **`float_model.summary()`** line displays an output of top and bottom layers. This will appear as the red text in the Jupyter Notebook cell. Just because it's red doesn't mean it's an error. It's worth looking at the top and bottom layers. Also, note the Output Shape changing as it passes through the layers (see Figure 10-9).

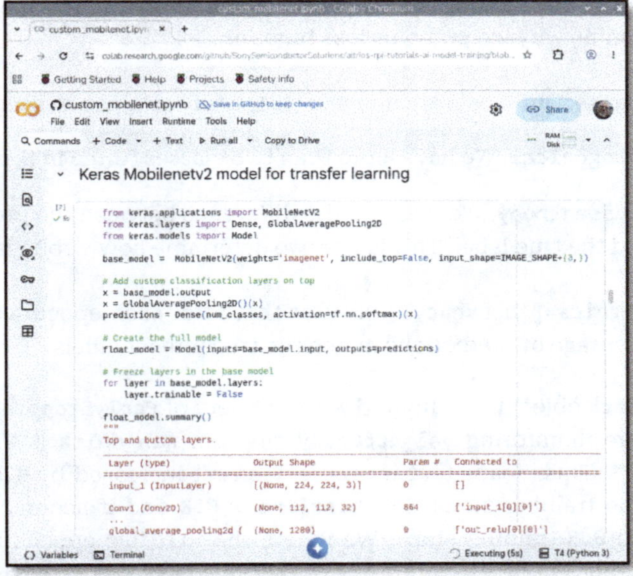

Figure 10-9   The layers our model will pass through to get to our final trained layer with three classification layers

Train

We are ready to roll. The next step is to train the model.

We start by setting the maximum number of **EPOCHS** to **20**. This is the maximum number of full passes we make over our training data. Typically, we find the training process stops at around 10–11 epochs when the **`EarlyStopping`** condition is met.

Our **optimizer** is Adam (Adaptive Moment Estimation). This works well with large datasets and complex models and uses memory efficiently. The optimizer is the algorithm that adjusts the weights in our neural network as we pass through the training data.

Our *loss function* is `SparseCategoricalCrossentropy()`. The loss function measures how wrong our predictions are. There are three elements that determine why we pick this loss function:

- **Sparse**. Our labels are integers (0, 1, 2 for rock, paper, scissors)

- **Categorical**. We have more than two classes (three in this case).

- **Crossentropy**. We measure the difference between the prediction and the true label. This is how we determine how wrong it is.

Setting `metrics=['accuracy']` ensures that after each epoch we report what percentage of predictions match the true integer labels.

Our **callback** object is configured with a variety of `EarlyStopping()` values. We are monitoring `val_accuracy`, and our `min_delta` is `0.01` and `patience=5`. This means that if accuracy hasn't increased by `0.01` for `5` epochs, the training stops. Our **baseline** is `0.8` and if accuracy never gets above `0.8` training stops. We ignore the first five epochs because these are quite noisy in nature. Next `restore_best_weights=True` rolls our model back to the epoch with the highest accuracy (not the final training epoch). Finally `verbose=1` configures the callback to display messages as the model is running.

Our `history = float_model.fit()` code fits everything together. The `fit()` method is Keras's training loop. We pass in `train_ds` (the training dataset), and `validation_ds` (the validation dataset), set `epochs` to our `EPOCHS` constant (20), and pass in the **callback** object we just created.

Figure 10-10 shows the training epochs in progress. Each epoch iteration improves the accuracy of our model.

Finally, we save the model to disk with `float_model.save(MODEL)`. Congratulations, you've just trained an AI model. Now we need to do something with it.

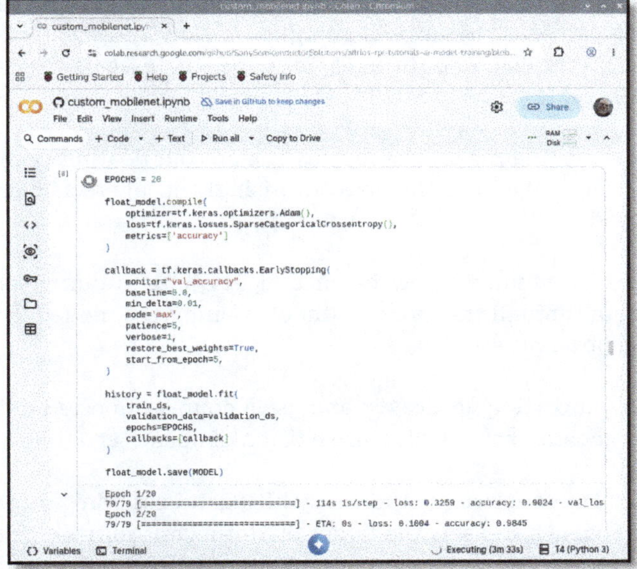

Figure 10-10   Training in progress

SEE THE FILE

Open the File browser (**View > File Browser**) and **models/mobilenet-rps** to see the **saved_model.pb** (protobuf) file. You can't open it, yet.

Quantization

Now comes the next important step. Quantization (**Figure 10-11**) is where we take the saved model and make it more efficient so it can run on our AI Camera. To do this we trade the floating-point numbers (Float32) into 8-bit integers (int8). It sacrifices a small amount of accuracy for substantial speed gains.

The floating-point values we used to train the model represent a huge range of numbers. A signed 32-bit floating point number represents over four billion values. This range is going to be slow and memory-hungry on our Raspberry Pi, so we shrink it down to 8-bit integers. These have a range of 256 values (-128 to +127).

This is what our quantization step does. It maps our floating-point values to integer values. We lose some (but surprisingly little) precision and gain a lot of speed.

It does this by running a few batches (`n_iter=10`, in our case) of real training data through the model without training the model. It forward passes and observes the ranges.

We start by importing `Generator` and use a couple of nested definitions to create a generator that will return a batch of images at a time.

Our next cell of code sets up the heavy lifting. It uses Sony's `model_compression_toolkit` and sets the target platform capabilities (tpc) to `tensorflow` with `imx500`.

The `q_config` value sets the configuration including the activation and weights error method; both use MSE (Mean Squared Error). We also set `weights_bias_correction` and `shift_negative_activation` to `True` and set the `z_threshold` to `16`.

Quantization is a deep subject and these values can be adjusted to improve performance. Take a look at Sony's `mct-model-optimization` GitHub page for more information (**rpimag.co/mctgit**).

Finally, we run our `mct.ptq.keras_post_training_quantization` code which returns a Keras model that we store in our `quantized_model` object alongside a `quantization_info` object.

Finally, we export it to our Jupyter Notebook file space with:

```
mct.exporter.keras_export_model(model=quantized_model,
                                save_model_path=MODEL_KERAS)
```

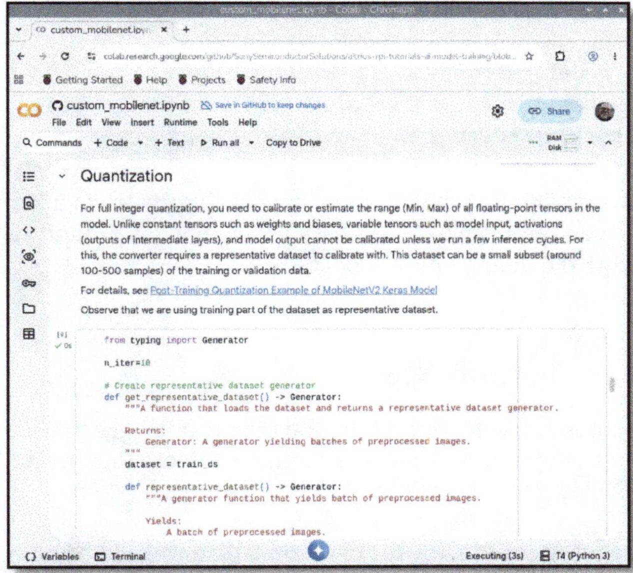

Figure 10-11   The quantization step makes our trained model efficient enough to run in Raspberry Pi hardware

Open the File Manager and you will find **mobilenet-quant-rps.keras** under the **models** folder.

Evaluation

We now move towards evaluating the model to ensure that it runs. We first ensure that our evaluation is set to run on the validation dataset:

```
float_model.evaluate(validation_ds)
```

As this runs it will output the loss and accuracy values. Typically, we want accuracy above 80%. This is our original Float32 model.

Then we run a **quantized_model.compile()** function and run the same evaluation against our **validation_ds** (validation dataset).

```
quantized_model.compile(
    loss=keras.losses.SparseCategoricalCrossentropy(),
    metrics=["accuracy"])
```

This is our compiled Int8 model, so we expect slightly lower accuracy, but in our case we went from 0.8306 to 0.8172 (we only lost around 1.5% of accuracy during the quantization process.)

Visualize detections

Our next step is to use Matplotlib to draw the test images in our database and overlay the predicted label and prediction score (how confident our trained model is that it is rock, paper, or scissors).

We load the test dataset with **split=["test"]** so we only bring in the images that our training model didn't see. We use **with_info=True** to ensure we get the class names.

Our **preprocess_image_visualization()** function resizes the image to the 224×224 size MobileNetV2 expects and we preprocess back from 0-255 and -1 to 1.

Our **detect_objects()** function gets the probabilities of the image. Remember it gets the probability for all three classes. It will look like **[0.01, 0.92, 0.07]** representing our image being 1% sure of rock, 92% sure of paper and 7% sure of scissors.

Next **get_top_prediction()** sorts the predictions and slices off the top indexed item. Our **top_class** variable is assigned the **class_name** (the human readable name).

Finally, our **visualize_detection** function displays the test image in Matplotlib and shows the class and confidence score. The result is a neat visualization of the images and results (see Figure 10-12).

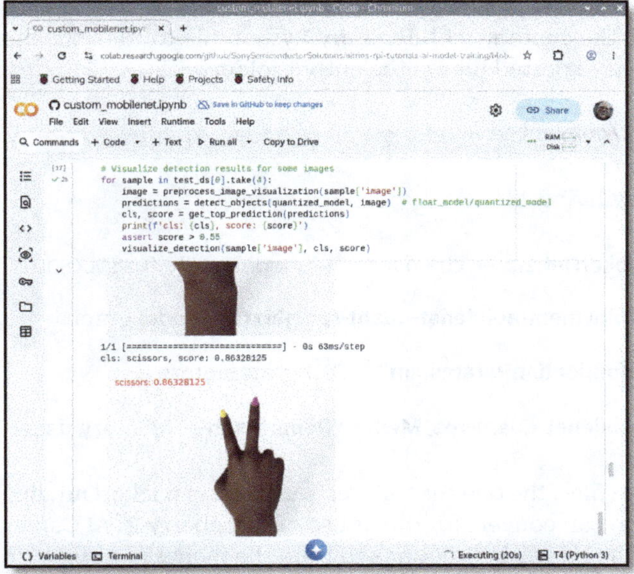

Figure 10-12   After the model is trained, the Evaluation and Visualize detections cells demonstrate that the trained model is functioning correctly

Conversion

Our final cell performs a relatively simple conversion and outputs the model to a **packerOut.zip** file.

```
!imxconv-tf -i {MODEL_KERAS} -o converted
```

As with our first **!pip** cell right at the start, the **!** lets us know that this is a system shell command. The **imxconv-tf** program was installed inside the Docker installation. It takes the Keras model we saved earlier and produces a **packerOut.zip** file which is compatible with our IMX500.

Finally, we use **!ls converted** to check all the files in our converted folder and an **assert** command to check that the **packerOut.zip** file is present.

Examine the files from custom_mobilenet.py

If you use the **custom_mobilenet.py** Python file to train the model, you will find the exported files in the output folder with:

```
ls output/converted
```

Here you will find:

- **packerOut.zip** — The packaged model ready for IMX500

- **deploymentmobilenet-quant-rps.pbtxt** — Model graph

- **definitiondnnParams.xml** — DNN parameters

- **mobilenet-quant-rps_MemoryReport.json** — Memory usage report

Of all these files, the one that interests us most is **packerOut.zip**. It is this file that we can convert for direct use on Raspberry Pi AI Camera. If you followed the earlier instructions to train the model in Google Colab, you should have already downloaded the **packerOut.zip** file to your **~/Downloads** directory.

Package the file

The final step packages the model into an RPK file. When running the neural network model, we'll upload this file to the AI Camera.

You do this on Raspberry Pi in terminal with imx500-tools. Install with this command:

```
sudo apt install imx500-tools
```

To package the model into an RPK file, make sure you've changed directory to the **custom_mobilenet** folder from your local copy of the GitHub repository. (If you haven't already cloned the GitHub repository, see *Example code* on page ix first.).

```
cd ai-projects-raspberrypi/ch10/custom_mobilenet
```

Next, run the following command, passing in your path to the **packer-Out.zip** file and output location:

```
imx500-package -i <path to packerOut.zip> -o <output folder>
```

If you used Google Colab to train the model and downloaded the **packer-Out.zip** file to **~/Downloads**, run this command:

```
imx500-package -i ~/Downloads/packerOut.zip -o .
```

If you used the **custom_mobilenet.py** Python file to train the model, run this command:

```
imx500-package -i output/converted/packerOut.zip -o .
```

This command should create a file named **network.rpk** in the **custom_mobilenet** folder. You'll pass the name of this file to your IMX500 camera applications.

> **IMX500 PACKAGER**
>
> Raspberry Pi's documentation covers packaging files for Raspberry Pi AI Camera: **rpimag.co/aicampack**
>
> Sony's AITRIOS website has an IMX500 Packager User Manual that outlines the packaging process in more detail: **rpimag.co/imx500pack**

Detect rock, paper, or scissors

We have provided a Python application called **rps_classifier.py** in the repository. Now that you have your **network.rpk** file inside the **output/converted** folder you can test the model out. If you haven't done so already, you'll need to install the software required to use the AI Camera. While you're at it, you should install **python3-opencv** and **python3-numpy**, which are two libraries needed to run the application:

```
sudo apt install imx500-all python3-opencv python3-numpy
```

Reboot when you're done with the installation, open a Terminal, change back into the **custom_mobilenet** folder and you're ready to test the model:

```
cd ai-projects-raspberrypi/ch10/custom_mobilenet
python rps_classifier.py
```

Make rock, paper, and scissor gestures and the preview window will display a bounding box around your hand displaying the gesture, as shown in **Figure 10-13**.

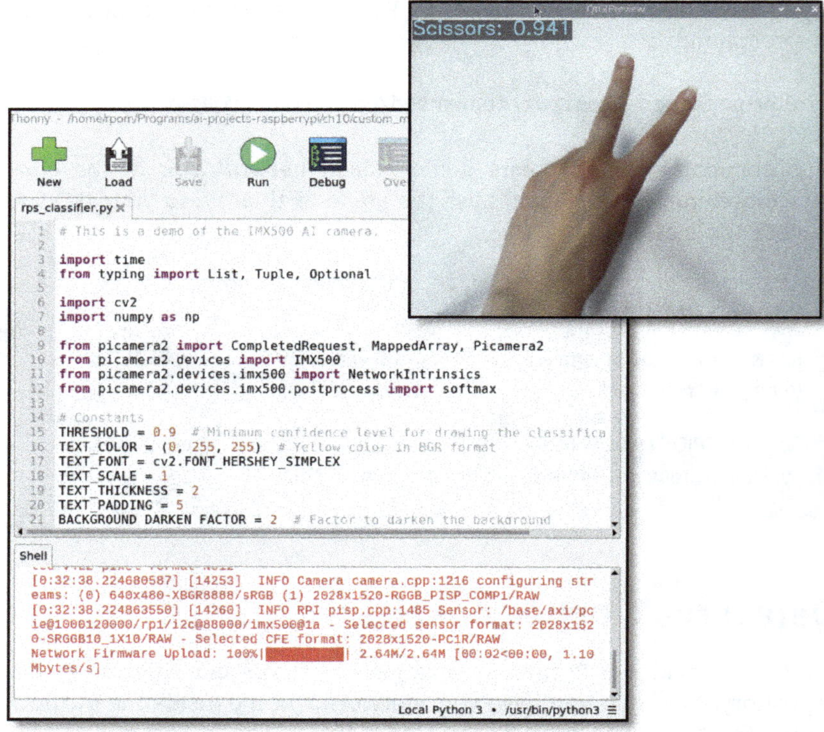

Figure 10-13 AI Camera detecting a Scissors gesture

Appendix A

Glossary of terms

Artificial Intelligence (AI)

An umbrella term for technologies that are expanding the capabilities of computers to include areas that were previously the domain of human intelligence.

AI model

A mathematical representation of a system that has been created by repeatedly training it on data.

Categorical value

A value that can't be represented as a numerical quantity. It can be a value that answers the question "what group does this data point belong to?" but can also answer yes/no questions.

Conditioning

Modifying a data set to prepare it for analysis. This can include transforming it by smoothing the data or dropping invalid data from a data set.

Containers

An alternative to a true virtual machine that lets you run applications in an isolated environment on your computer. They use your host computer's operating system kernel but have their own versions of runtime libraries and supporting applications.

Context

The total amount of information that you have supplied to a model, typically in an interactive session such as a chat. Context is typically measured in number of *Tokens*, and models have a limit to how many tokens they can handle before exhausting the available context. The context determines how much the model can remember of your chat.

Continuous value

A value that can be represented as a numerical quantity and can take on any value within a given range of values.

Convolutional Neural Network (CNN)

A type of *Neural Network* designed primarily for processing grid-like structures, in particular images.

Dependent variable

Data that you want to predict. See also *Label*.

Diffusion model

A *Generative AI* model that can create realistic images, audio, or video. They are trained by adding noise to images with Gaussian noise. They then run the process in reverse, gradually predicting and removing noise from an image until it looks or sounds like something humans would recognise.

Feature

Another word for *Independent variable*.

Generative AI

A field of artificial intelligence that develops models capable of creating new content (images, text, video, etc.). These models are trained on enormous quantities of data, and produce content in response to textual prompts, such as "a human-sized cat driving a bus."

Independent variable

Data that you already know. See also *Predictor*.

Inference

Running an AI model on new data to generate predictions, classifications, or decisions.

Label

An answer to a question such as "if the humidity is 100% and the cloud cover is 100%, is it raining?" In supervised learning, training data is labelled with a value that is known to be correct.

Linear regression

A foundational modelling technique that can predict the relationship between one or more *Independent variables* and a *Dependent variable* that has *Continuous values*. It is used when changes in one or more variables are associated with consistent changes in another. Predictions are made with a straight line in which points along the y-axis represent the numeric outputs of the model.

Logistic regression

A foundational modelling technique that is similar to *Linear regression*. However, in logistic regression, the *Dependent variable* has *Categorical values*. Predictions are made with an S-shaped curve that represents probabilities from zero to one.

Matrix multiplication

A mathematical operation performed on matrices of data. These are particularly useful for neural networks. Graphics cards or dedicated AI hardware perform these operations more quickly.

Model

A representation of a system that learns from data.

Neural Network

A computational model inspired by the connections in the physical brain. Neural networks are built of layers of mathematical functions called *neurons* (or nodes). Each neuron takes an input (from a connected neuron), multiplies it by *weights and biases*, sums the result, passes it through an activation function, and produces an output.

Neural Processing Unit (NPU)

A hardware processor, such as the Hailo-8L, Hailo-8, and Hailo-10H, designed to accelerate *Neural Network* calculations.

Overfitting

When a model's results correspond so closely to the training data as to be useless with real-world data.

Predictor

An input variable in your training data that predicts a particular value for the label. In the example given earlier for the *Label* glossary entry, the predictors would be humidity and cloud cover.

Quantization

A process of optimization that maps larger values into a smaller set to reduce a model's complexity and memory usage.

Regular expression

A syntax for matching patterns within text. For example, the regular expression `\d\.\d` would match a single digit followed by a decimal point and another digit (such as `1.0`).

Supervised Learning

A machine learning technique where every combination of *Predictors* in your training data has a corresponding output value (the *Label*).

Tensors

Multidimensional arrays of data used in AI models.

Test data

A subset of your training data used for evaluating the outcome of the training process, but not used in the training process itself.

Token

A representation of a meaningful chunk of text. This includes spaces and special characters. It also includes metadata tokens such as the start or end of a sentence.

TOPS (tera operations per second)

A metric for measuring the speed of AI hardware that represents one trillion operations per second.

Training

The process of creating an AI model by exposing it to data repeatedly and measuring the output while adjusting the weights and biases.

Unsupervised Learning

A machine learning technique where the training data is unlabelled (see also *Labels* and *Supervised Learning*). Because you don't have labels, the model needs to look for patterns in the data on its own.

Validation data

A subset of your training data used for adjusting the training process but not used in the training process itself.

Virtual environment (Python)

A Python sandbox for installing modules without affecting your operating system's primary Python installation.

Weights and biases

Parameters within the AI model that are adjusted during training to improve accuracy.